D1351631

Slipform Concrete

R.G. Batterham

Slipform Concrete

R. G. Batterham

THE CONSTRUCTION PRESS
LANCASTER   LONDON   NEW YORK

LIBRARY

Limerick Technical College

CLASS NO. : 624.1834 BAT

ACC. NO. : 3346

The Construction Press Ltd,
Lunesdale House, Hornby
Lancaster, England.

A subsidiary company of Longman Group Ltd, London.
Associated companies, branches and representatives
throughout the world.

Published in the United States of America by
Longman Inc, New York.

© R. G. Batterham, 1980

All rights reserved. No part of this publication may be
reproduced, stored in a retrieval system, or transmitted
in any form or by any means, electronic, mechanical,
photocopying, recording, or otherwise, without the prior
permission of the copyright owner.

First published 1980

British Library Cataloguing in Publication Data

Batterham, R G
  Slipform concrete.
  1. Concrete - Slipforming
  I. Title
  624'.1834   TA682.44

  ISBN 0-86095-855-8

Printed in Great Britain by
William Clowes (Beccles) Ltd, Beccles and London.

To my sister Elizabeth, for typing the manuscript
and for her encouragement and patience.

ACKNOWLEDGEMENTS

The author and publishers wish to thank the companies and institutions below for permission to use diagrams, graphs, tables and photographs of which they hold the copyright.

Barcu Construction Co. Ltd., Figures 1.7, 5.3, 5.4, 5.5, 5.17, 5.18, 5.19, 5.20, 5.24; Brady & Partington (Consultants); British Lift Slab Limited, Figures 1.3, 2.4, 2.5, 2.6, 2.8, 2.10, 2.11, 2.12, 4.1, 5.25, A6, A7, A8; Cement and Concrete Association, Figures 3.1, 3.2, 3.3, 3.4, 3.5, 3.6, 3.7, 3.8, Tables 3.1, 3.2, 3.33; Chesterfield College of Technology; The Concrete Society, Figures 2.1, 2.2; Don Valley Engineering Company Limited, Figure 5.1; Douglas Technical Services Limited; John Laing, Figures 2.9, 5.7, 5.8, 5.11, 5.16, 5.21, 5.22, 5.23, A17, A18, A20, A21; National Coal Board, Midlands Regional Laboratory, Figure 5.6; National Coal Board, North Derbyshire Area, Figure 5.1; Ready Mixed Concrete (East Midlands) Limited; H. W. Rhodes & Partners (Consulting Engineers) Limited, Figures 5.1, 5.2, A1, A2, A3, A4, A5; Spectrum Photos of Chesterfield, Figure 2.7.

Contents

Introduction

Slipforming has been accepted as a succinct construction technique within a comparatively short period of time — an indication of its considerable popularity. This popularity has in turn encouraged further research, ensuring the adoption of new methods and modern materials which have firmly established slipforming as an economical, rapid and accurate form of construction.

Its origin and history remain something of a mystery but it is generally accepted that it first appeared in 1885 when a Texan named Carrico used the principle to build a small concrete shaft. No further development was apparent until 1903, when the Americans first used a screw jack to propel the formwork. This was the first true slipform system, for, although manually operated, it showed up many of the problems of concrete vibration, fallouts, honeycombing and wall damage by buckling climbing-tubes which had to be overcome to establish the concept. During the 1940s, a Swedish manufacturer developed the now-standard hydraulic equipment which enabled men such as Jesperson to produce records in the 1960s, when he constructed a chimney 4.50m in diameter and 32.40m high in nine days.

Now that it is properly developed, the technique can be applied to many different forms of structure, including tapering formations with straight or parabolic profiles incorporating constant reductions in wall thickness. Traditional applications for slipforming were silos, simple chimneys and the like, but common applications now are the construction of building cores, slender bridge-piers, chimneys and water towers.

Although systems operated in the Western world make extensive use of hydraulically operated equipment, with standard units creating the sliding apparatus, manually controlled systems may still be found operating in India, Central and Southern America.

Section through screw jack sliding formwork system.

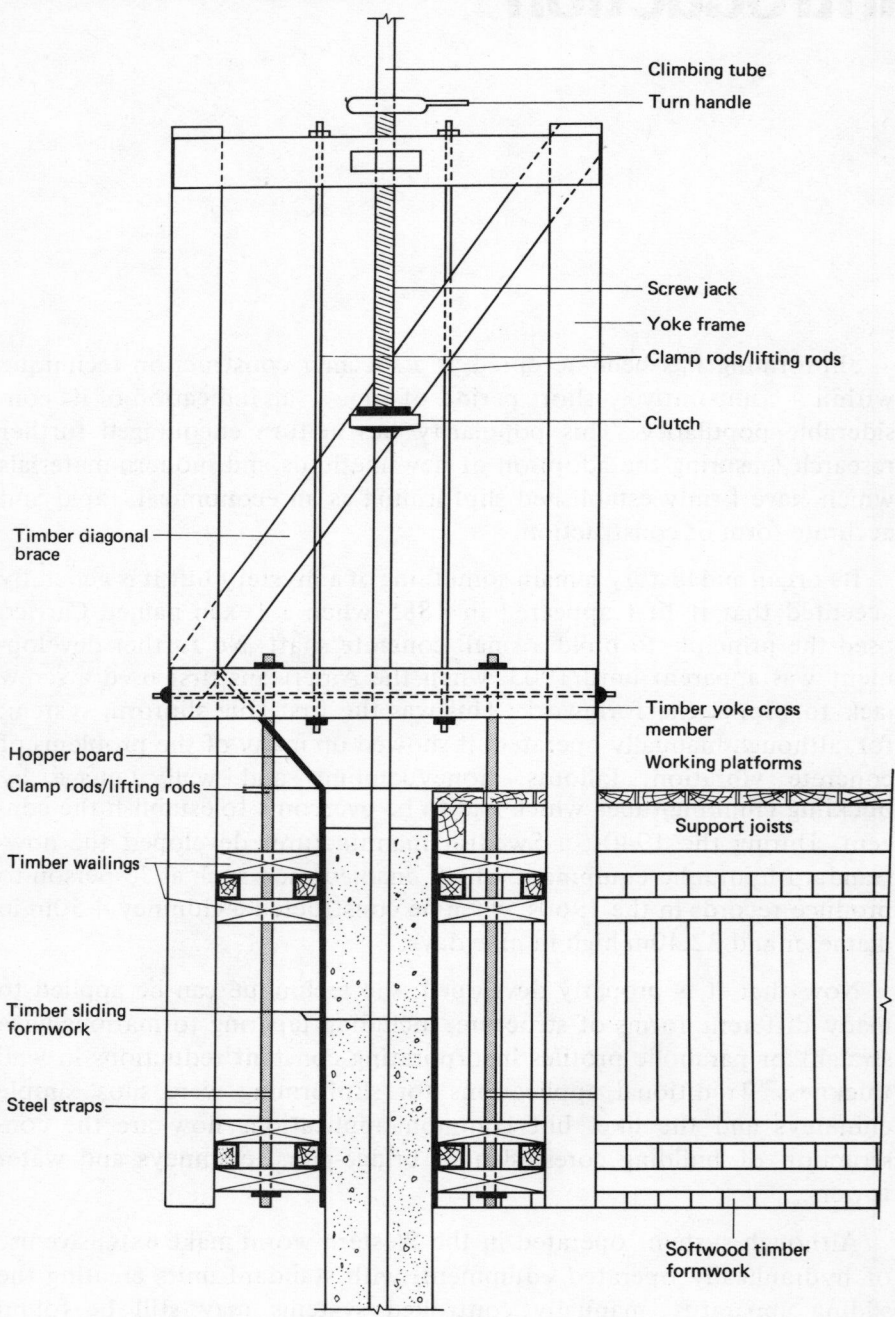

Climbing tube

Turn handle

Screw jack

Yoke frame

Clamp rods/lifting rods

Clutch

Timber diagonal brace

Timber yoke cross member

Working platforms

Hopper board

Clamp rods/lifting rods

Support joists

Timber wailings

Timber sliding formwork

Steel straps

Softwood timber formwork

Normally, slipformed structures are inherantly stable due to their design, but certain slender formations may require bracing until floors and other sections are completed. Theoretically it is possible to arrive at an unstable situation where the rate of construction far exceeds the rate at which the concrete gains its strength.

The economics of slipforming compared with competing techniques depend largely upon the type, size and complexity of the structure under consideration. In Great Britain for example, any single formation more than 30m high can be considered a prime candidate for the technique, though where a series of smaller structures are to be undertaken this figure may be significantly reduced. To ensure the rapidity, accuracy and economy of the system are not destroyed, extreme care in organising the supply of labour and materials is of prime importance.

As an introduction to the nuts and bolts of the technique, let us take a closer look at the screw-jack method, as applied to a tapered chimney (see illustration).

Curved inner and outer formwork faces were made from manufactured tongue-and-groove boards secured to timber ribs. This arrangement, supported by timber yokes, was all raised together on climbing tubes embedded in the structure's concrete footings. Even at this stage of slipform's history it was discovered that it was best not to incorporate a concrete kicker to start the structure on its way. However, to prevent the important initial grout from escaping underneath the shutters, damp sand was packed around the formwork base.

Both formwork faces had to be fitted with adjustable devices to enable the chimney's circumference to be continuously reduced. These adjustable units were manufactured from steel plates, each with one end fixed to the formwork while the other over lapped its opposite number. A turnbuckle arrangement was used to slide the plates across one another, thus reducing the formwork's circumference. It should be noted however, that both formwork faces had to be adjusted simultaneously in order to preserve a continuous wall thickness. Naturally, corresponding adjustments were also needed to the yoke assemblies in severe situations.

To prevent the whole system from rotating due to the horizontal force applied to the jack handle, alternate jacks used a left-hand thread, so that each counterbalanced the horizontal movement of its neighbour.

Stability was maximised by staggering the reinforcement joints; in the lowest tier of vertical steel reinforcement, each successive bar on the structure's circumference was longer than its neighbour, the bars' upper ends thus forming a spiral. This spiral could be maintained as the structure rose simply by using bars of equal length for all successive tiers, apart from the top, where the first-tier length progression was reversed to leave all the ends at the same height.

The lifting operation was carried out by a number of people, the figure depending upon the size and shape of structure. Normal procedure was to use a colour code, with each operative being responsible for a certain number of specially coloured jacks. When a whistle was blown, these were rotated by a quarter of a revolution to lift the whole structure uniformly. The colour coding practice ensured that no jack was missed.

On larger contracts a permanent levelling hut was often constructed on the uppermost deck, from which levels could be frequently checked. It was also common practice to fix sight rails at each yoke position so that engineers could carry out individual checks on both jacks and yokes.

Once sliding was completed, the formwork was pinned to the walls, so providing greater stability during the dismantling operation. If a concrete roof was required, the climbing equipment could then provide the staging from which to support the necessary formwork. In such instances a collapsable working platform was used to facilitate dismantling of the internal formwork. Apart from this complication, striking formwork was a straightforward operation commencing with jacks and yokes being boxed out, then formwork materials being removed and lastly the decks and yokes being lowered to ground level.

1 Equipment and materials

The acceptance of the moving-formwork concept by the construction industry paved the way for a rapid growth of simple silo and chimney formations. However, as knowledge, experience and technical backing increased a wider variety of structures became candidates for slipforming.

Unfortunately, being a specialised form of construction, each new contract is confronted by the perpetual problem of technical ignorance, throughout the whole strata of site staff. To overcome this problem, the specialist equipment suppliers also provide all the necessary supervision and skilled operators.

The slipform system of sliding formwork can be broken down into six main components and, although the original concept of moving formwork remains the same at heart, the component materials are regularly updated to suit specialist requirements and secondly in an attempt to overcome inflation.

Formwork

The initial formwork configurations were created from a variety of assorted timbers, typically 25mm thick and 1.2m deep. Moreover the actual face was constructed from a series of tongue-and-grooved boards which also proved to assist in controlling any lateral swelling due to their constant contact with wet concrete. Expansion gaps were incorporated within the formwork face to provide for controlled movement (see Figure 1.1).

Generally therefore, the timber forms were very rigid but in order to obtain a suitable finish, various precautions had to be taken, including a batter of about 6mm which was incorporated within the completed formwork. This extra width, usually at the base of the formwork, prevented the concrete from adhering to the timber surface.

13

Figure 1.1 Elevation of a typical timber formwork panel.

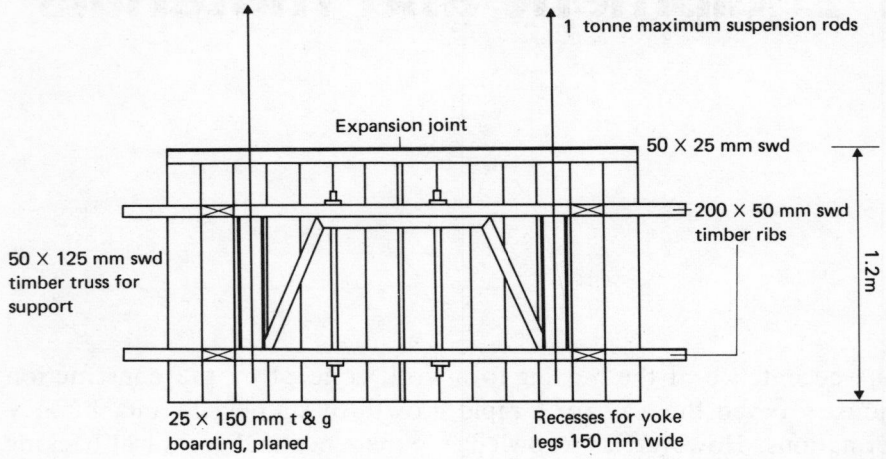

1 tonne maximum suspension rods

Expansion joint

50 × 25 mm swd

200 × 50 mm swd timber ribs

50 × 125 mm swd timber truss for support

1.2m

25 × 150 mm t & g boarding, planed

Recesses for yoke legs 150 mm wide

Detail of the expansion joint

Tongue removed and groove cut

Originally the formwork panels were suspended by long steel rods from yokes about 1.8m apart, the actual distance depending on the panel size. This system, however, required an excessive degree of bracing and today the now steel or glass fibre panels are instead fitted directly to the yokes. Such panels are expensive to buy but can be used repeatedly, give a high degree of finish and are immune to damage by poker burns.

Once the formwork has been assembled, treated and checked for alignment and level, it is tightened up and filled with concrete. Today's hydraulic jacks have largely eliminated the problem of formwork rotation referred to in the introduction, but if it does occur, special pulling tackle can be attached to the reinforcement and formwork to rectify matters.

Ribs

Traditionally these have been formed from timber but rising costs are favouring the adjustable steel rib, which can be used again and again. Two rows of timber ribs were used on the traditional systems, bolted and braced together to form a truss which transmitted the loads directly to the yokes (see Figure 1.1).

14

Figure 1.2 Third and lowest deck of a Siemcrete system.

Timber ribs, usually notched to clear the yokes, were formed from 100 x 150mm or 100 x 200mm sections for straight formwork panels and from 50 x 100mm or 50 x 200mm sections for circumferential panels.

The modern steel rib carries out four functions:

(a) to support and hold the forms in place,

(b) to support the working platform,

(c) to support the suspended platform,

(d) to transmit the lifting forces from the yokes to the formwork.

Yokes

Here again steel has replaced timber, channel section now being the normal material. However, the yokes still serve the same two functions:

(a) to transmit the lifting forces from the jacks to the ribs,

(b) to hold the whole system together in a given pattern.

The horizontal member of the yoke is always used to house the jack while the legs of the yoke hold the formwork, ribs and decking in place (see Figure 1.2).

For safety reasons, any couple which is created within a yoke must be restrained. The necessary horizontal thrust can be arranged by passing a series of 30mm diameter steel tie rods across the centre of the whole slipform structure. These rods create a feature known as the spider and ensure that the thrust is taken by the deck or by special thrust timbers incorporated within the deck system.

Working platforms

The working platforms on both present-day and early slipform systems use timber extensively for decks and supporting joists. Although timber may be replaced by other materials in certain situations, in general nothing as yet has offered the same versatility in relation to the working platforms.

The early slipform systems began with two decks, the top one being designed as a working deck and storage area and the lower deck for finishing the concrete as it emerged from the formwork. Later it became apparent that the top deck was really too small for both storage and working purposes — hence the addition of a further deck higher up.

Figure 1.3 Typical details of Siemcrete one slipform system.

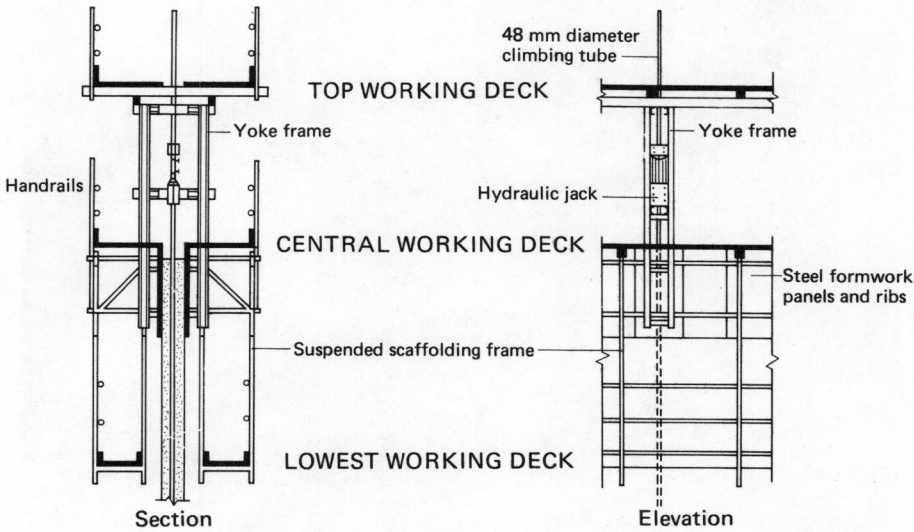

This additional deck is used to store materials and for unloading the concrete from the crane (see Figures 1.2 and 1.3).

The weight of material carried on the working and storage decks should be checked frequently, as excessive loading will exert undue forces on the yoke structure, which may move out of line as a result.

As an aside, it is worth noting that when a flat roof is to be incorporated the main working deck can be designed to provide a base to support the roof formwork. This technique needs consideration and calculation however.

Jacks

Three types of climbing jacks are available — screw, hydraulic and pneumatic — and the latter pair operate by anchoring the extended jack head to the climbing tube (or bar) and then pulling up against the anchorage to raise the yokes. Once the jack movement has been exhausted, the jack head must be repositioned further up the climbing tube — hence the use of a clutch to provide the anchorage. The climbing tubes or bars may either be withdrawn or cast in the concrete structure depending on cost, thickness of the walls (very thin walls could be damaged by the withdrawing action), wall reinforcement, wall insert details and lastly upon the time factor.

Figure 1.4   Details of the hydraulic jack arrangements.

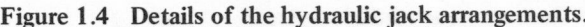

When designing a slipform system, a job which must be done for every contract, it is vital to ensure that the number and distribution of jacks are sufficient to lift the system without causing any undue stress to any part of the framework. If unwanted stresses occur, uniform upward movement will not be obtained and a distorted finish will result. The load carried by each jack is largely determined by three factors:

(a) dead weight of the whole assembly,

(b) the friction of the forms relative to the concrete,

(c) the area of working platforms and liveloads imposed.

An overloaded hydraulic or pneumatic jack will simply jam in position but an overloaded screw jack can be dangerous, as part of its locking pins may shear. With timber-based systems, cross-sections of the formwork area one jack may lift are as follows:

(a) small circular silos 5.6m²

(b) 12.2m diameter silos 5.1m²

(c) awkward shaped plan . . . . 3.7—4.2m²

(d) single faced formwork only 2.0m²

(e) general situations . . . . 3.1—7.3m²

The approximate load hydraulic jacks can lift without any undue concern is two tonnes, a figure confirmed by years of experience.

The early screw jack
This held a solid body working against a 25mm diameter steel climbing tube. No longer used, it was made up from an upper threaded portion 37mm in diameter, 1.2m long and with three square heads within the first 25mm. The lower shank was 22mm in diameter, 1.05m long in one section and having a square slot cut in its circumference which travelled the length of the bar.

The head cast with four arms bored to take a 19mm-diameter bar with a hole in the centre of 37mm, ensures that it will slip over the jack. Hence the head is free to move vertically but cannot move horizontally without the actual joint turning. The jacks shank, a 25mm-diameter pipe 900mm long, is inserted within the jack and the pressure of the jacks shank is transmitted by the collar at the junction of the thickened portion with the shank. As the headpiece is turned with the bar, the jack, which is working against the climbing tube, will raise the slipform framework. With this form of jacking system, the maximum rotation a jack is allowed per movement should not exceed a quarter of a revolution.

The hollow screw jack

This may still be found in operation around the world today. Typically, it has a 26 mm-diameter hole down the centre through which passes a 25mm-diameter climbing tube. The jack grips the tube through a clutch consisting of a series of screws and toggles which lock when the jack acts downwards. As with modern systems, the climbing tubes run through the wall centre and once the climb is completed the tubes may be withdrawn or if hollow filled with grout. The hollow screw jacks are 900mm long with a capstan head into which a 50mm diameter steel rod may be inserted to turn the jacks. The capstan head, 1.5m above the top of the deck at its highest point and 1m at its lowest, ensures an effective travel of 0.5m.

Hydraulic jacks

A steel climbing tube 48 or 30mm in diameter, set on the structure's concrete base, passes through the centre of the jack which in turn is attached to the yoke. Hydraulic pressure is applied to a series of grips within the upper element of the jack, and when activated these grips clamp directly onto the climbing tube. The lower element of the jack is then drawn up, lifting the whole slipform structure. Once the pressure within the hydraulic system is released, the upper element of the jack automatically climbs the tube in preparation for the next lifting cycle, which commences with the re-application of hydraulic pressure (see Figure 1.4). Uniform movement in this system is achieved by connecting all jacks to a central hydraulic pump. Hand-operated valves, located at various key points in the pipework, allow individual jacks to be operated or repaired.

Hydraulic jacking is versatile enough to permit either solo operation or, if conditions are suitable, complete automation. Each lift of the jacks only moves the formwork around 70mm, so the system moves in a series of jerks rather than with a continuous sliding action. The effect this can have on finish is shown in Figure 1.5. Attempts have been made to smooth out the process by incorporating electric motors within the jacks, but several problems arose, namely:

(a) concrete working its way into the electric motors,

(b) rough handling of the motors,

(c) sensitivity to overloading,

(d) synchronisation of the motor speeds.

20

ACC No· 3341·   624.1834
                                  BA7

Figure 1.5   Results of slipforming's sliding action on the concrete face.

Figure 1.6 In-situ climbing tubes.

Over the years climbing tubes have diminished in diameter from 50 to 30mm and are produced largely from mild steel sections in convenient lengths with carefully prepared interlocking joints. As described in the introduction, other lengths may also be used to form a rigid framework incorporating staggered joints (see Figure 1.6).

Spiders

Early slipform systems incorporated spiders which were manufactured from beams of timber and which interlaced the internal formwork ribs to provide what is now regarded as an excessive degree of rigidity. Present-day spiders, which offer little support to the framework by comparison, are produced from 6–30mm-diameter mild-steel bars attached at the outermost end to the steel ribs and at the inner end to a

Figure 1.7 Steel spider with central ring.

central plate. Complicated formwork often requires intricate intercon-
nected spider patterns; moreover, a second spider may be needed in
order to transmit forces within the three-deck system (see Figure 1.7).

Concrete design

Considerable research has been conducted into the suitability of
materials to develop an acceptable mix design. However, concrete design
is a complex subject in its own right and space does not permit a proper
treatment of the subject here. Suffice to say that these designs are
varied in accordance with changing temperature, required surface tex-
tures and performance of available cements.

As a rule the mix design should be tested before the slide commences
to establish the strength, workability, slump and initial setting times
and further tests should be made throughout the operation. Moreover,
special additives may be essential to attain a satisfactory performance
with a mix containing a high proportion of fine materials. Before leaving
the subject, it is worth noting that the ratio between concrete mix
design and atmospheric temperature determines the velocity at which
slipforming may be executed.

The case for slipforming

At this stage, having covered both the background and the principles of
slipforming, it is worth setting out the pros and cons of the technique.

Advantages

1. Accuracy
2. Continual casting, creating a monolithic structure
3. No joints unless a halt occurs
4. Lends itself to almost any shape in plan.
5. Strength
6. High quality surface finish
7. Rapidity
8. Labour saving in long term.
9. Saves formwork materials
10. Economical, for structures above a certain size

11. Produces aesthetically pleasing structures

Disadvantages

1. Good co-ordination and site organisation required.

2. Large quantities of equipment (eg, generators, lighting systems hoists) required.

3. Day and night shifts must be organised.

4. Labour force may require familiarisation with equipment and methods.

5. Operations must be continued during unsuitable weather.

6. High initial expense.

7. Need for 24-hour service facilities (eg canteen, material supply, maintenance crews).

8. Communication must be co-ordinated between ground level, crane drivers, slipform and hoist operators.

9. Social problems created by long working hours.

10. Fixing door and window frames etc.

11. Labour requirements before, during and after the slide, (see Chapter 5).

12. Less site control due to sub-contract labour.

2 Representative slipform ventures

When slipforming was in its infancy, the decision whether or not to use it in preference to other forms of construction often depended on the answer to the question 'is it technologically possible to use it on this job?' Such is the system's present-day sophistication that the modern designer is much more likely to ask, 'is it economic to use it on this job?' Technical feasibility is almost taken for granted.

To illustrate the complexity and variety of structure to which this method of construction can be applied, let us examine three quite different examples of slipforming.

Reading's new town centre (1974)

Occupying a 1.4 hectare site, this construction incorporated a 13-storey tower block, 46m high. Its central core measured 13.7 x 15.2m and housed five lift shafts, two staircases and various service ducts. The development now provides 18 000m² of office space, 9000m² of shopping facilities and a multi-storey car park for 1000 vehicles.

Erection and demolition of the three-deck slipform system took six weeks; the actual slide lasted only eleven days and nights (see Figure 2.1).

Equipment included hydraulic jacks and 25mm-diameter climbing tubes which were withdrawn on completion. Formwork and ribs were constructed in timber, 150 by 75mm for the ribs and 83 by 29mm kerwing hardwood strips for the sliding timber face. Once the formwork and sliding equipment had been installed, a test sample of the formwork in three demountable sections was erected and filled with concrete in 200mm-thick layers. At periods of one, two and three hours after filling, each section was removed independently to determine when the actual slipform framework could commence sliding — determ-

Figure 2.1 Reading new town centre, work progress graph.

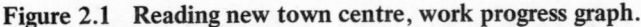

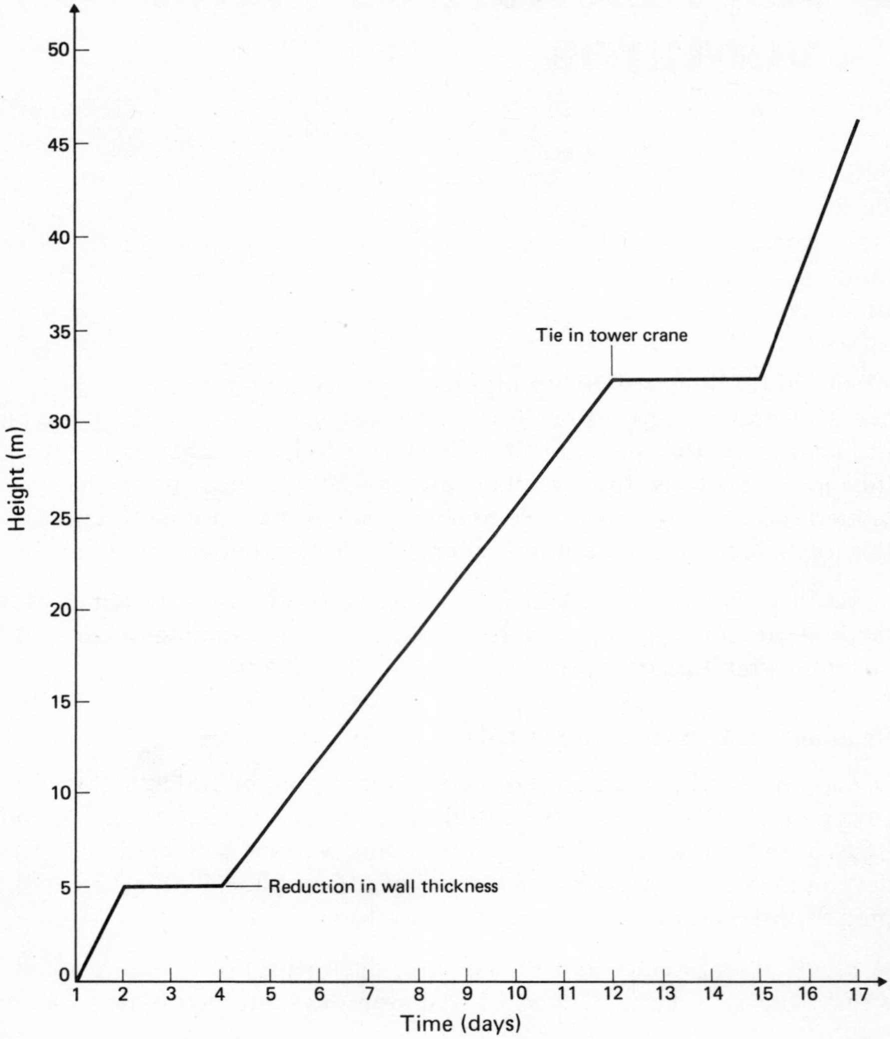

ined in this case to be three hours. Another test, which involved pene-
trating the fresh concrete from the formwork's leading edge with a
12mm-diameter steel bar until resistance was felt, determined the speed
at which slipforming could be carried out (see Figure 2.2).

Because of the five lift shafts in the central core, special precautions
were necessary to ensure absolute plumbness. Optical plumbs and
plastic targets were fitted to the slipform framework with their centres
coinciding and readings were taken every three hours. In addition each

jack incorporated a water level, the datum and subsequent measurements being taken from the tower crane's mast.

London's new Stock Exchange (1967)

This structure was designed round two main central cores 97.5m high, which were constructed independently and consecutively. Slipforming took seven weeks for the two structures, working day and night. The requirement for an early completion date prompted the decision to use slipforming, a decision which was made two years before the contract commenced. Being an intricate formation with many openings and variations in wall thickness (from 178mm to 914mm), close co-operation between all personnel was of the utmost importance, both while formulating and during the operation.

With the site so close to London's centre, material supply co-ordination was of prime importance. Acute traffic and site storage problems were relieved by having fixed material delivery times during the crane's idle periods, which were arranged to coincide with the quietest times in the surrounding streets.

Figure 2.2 Graph of level of stiffened concrete relative to formwork depth, Reading town centre contract.

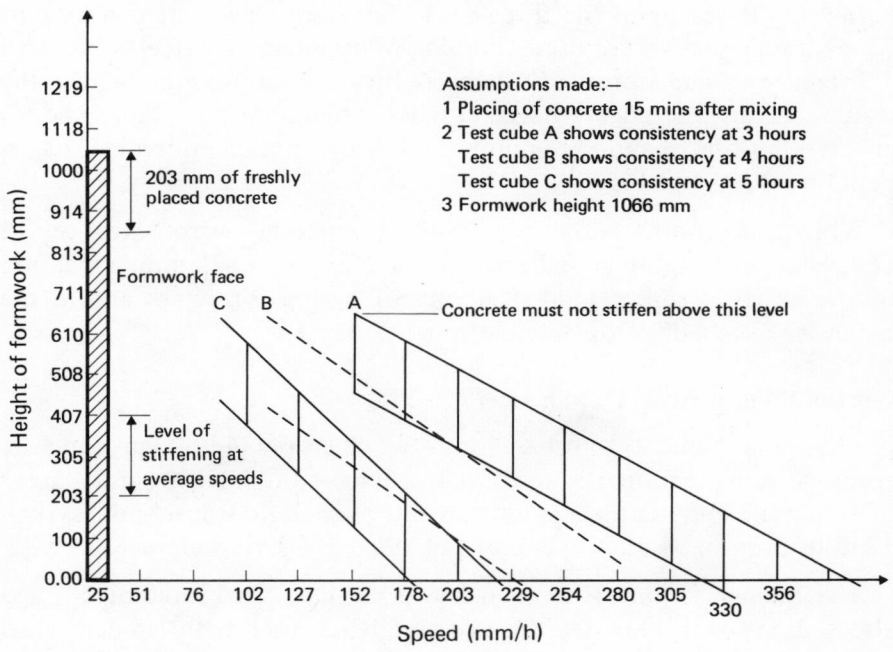

29

Figure 2.3 Yoke members, central deck and steel reinforcement relationship details.

Slide speed varied from 150 to 380mm/hr, depending upon weather conditions, steel reinforcement-fixing rate and production ability of the concrete plant, which under normal conditions was manufacturing 8.4m³/hr. Based upon the three-deck Siemcrete One system, a total of 75 hydraulic jacks were used climbing 50mm-diameter steel tubes with 1.2mm-deep rigid formwork panels. Fitted over 600mm above the leading formwork edge, the steel yoke cross-sections were fixed to ensure that the maximum number of horizontal reinforcement bars could be fitted (see Figure 2.3).

With this contract the maximum permissible error was set at ±63.5mm; in fact the actual error did not exceed ±12.5mm, a measure of the extreme care extended in controlling the formwork and of the extensive use made of optical instruments.

Central Bank offices, Dublin (1977)

This development was based on two 8-storey 43m-high concrete cores, of cross-section 12.6m by 8.1m. Each contained the usual array of stairs, lift and ventilation ducts, while at each floor level no less than 12 door frames, 14 duct openings and 19 steel inserts were necessary.

The Siemcrete One system, using an average workforce of 24, was also chosen for this job. On this occasion it was used with 1m-deep steel formwork panels to create the 300mm outer and 200mm inner walls. A system of 48 hydraulic jacks provided the movement, climbing 50mm-diameter steel tubes at 2.5m centres at 70mm per stroke. This gave a target speed of 250mm/hr, with the maximum and minimum speed determined as 450 and 150mm/hr respectively. The target speed can be translated into one wagon load of $30N/mm^2$ readymixed concrete per hour, a rate which proved easy to cope with on the top deck whence it was transported by crane and skip.

Accuracy of ±20mm was achieved by using optical plumbs, assisted by fixing steel tapes at floor and central deck levels. The tapes allowed a series of horizontal marks to be made on the climbing tubes above the top deck every 350mm, so providing levels for jack adjustments.

Figure 2.4   Europa bridge, Austria. Supported by concrete piers, the bridge's height is 147m. Each pier has four tapered sides. *(Licensee: Universale, Vienna)*

Figure 2.5 Stock Exchange building, London. Designed with two independent cores which were slipformed consecutively, this structure is 97.5m high. *(Licensee: British Lift Slab Ltd)*

Figure 2.6 Chimney in Burton-on-Trent, England. This unusual shape chimney semi-oval in plan, is 48m high and has fluted side features. *(Licensee: British Lift Slab Ltd)*

Figure 2.7   Coal storage bunker, Bolsover, England. A circular shaped structure, the bunker incorporates two changes in wall thickness and a traditionally constructed concrete ring. The overall height is 36m. *(Licensee: British Lift Slab Ltd)*

Figure 2.8 Bund wall, Cork, Eire. With an internal diameter of 37.5m and a height of 23.5m, this structure was wire wound on completion. *(Licensee: British Lift Slab Ltd)*

Figure 2.9 Administrative block, Tameside District Council, Ashton-under-Lyme, England. This building has a single independent core and internal walls. *(Licensee: British Lift Slab Ltd)*

Figure 2.10 Telecommunication tower, Reisenbach, West Germany. This tower comprises of a main shaft 126m high and an aerial shaft 50m high, the latter section being cylindrical while the lower shaft section is tapered. *(Licensee: Siemens, Bauunion, GmbH)*

Figure 2.11 Silo system, Ladybrand, South Africa. The machinery tower was slip-formed to a height of 42m and each of the surrounding circular storage bins to a height of 31m. *(Licensee: Concor, Johannesburg)*

Figure 2.12 Standard Bank building Johannesburg, South Africa. The 100m-high central core contains three intermediate supports of pre-stressed cantilevered beams from which nine pre-fabricated floor slabs are suspended. *(Licensee: Concor, Johannesburg)*

3 Main design principles for sliding formwork

Introduction

The essential feature is to design a system which is strong enough to cope with the pressures created. If this part of the job is done thoroughly, several beneficial by-products result: high dimensional accuracy, good cost control and an improved structure generally. Essentially the design of slipforming formwork revolves around three factors: dependable loads, the friction between concrete and formwork and the pressure exerted by the concrete on the formwork.

However, the lateral pressures exerted by freshly placed concrete on the formwork, especially when the latter is stationary, are enormously and unpredictably variable. This situation is aggravated by the effect which the inward battered formwork faces have on the setting concrete as they move upwards. The pressures are a result of concrete consistency, weather conditions, type of formwork face, wall thickness, rate of concreting, type of compaction, initial set and amount of reinforcement.

Methods of calculation

Pressure and friction factors have to be considered most carefully and may be split into two groups depending on their source.

Group number one:

Effect of design, technology and special features liable to alter during the slide. (eg, wall thickness, slide speed, initial concrete set, nature of formwork faces, concrete compaction and concrete consistency).

Group number two:

Effects of the outside factors (eg, live loads, difference in travel of lifting gear and wind pressure).

Group one factors, although variable, are controlled and known from the start. Assuming the slide is slow, the formwork batter will cause separation from the concrete at distance h_1 (see Figure 3.1). On the other hand, if the slide speed is excessive the formwork will separate at h_2. Hence h_1 or h_2 provides the contact height at which the effective pressure head of the concrete acts on the formwork. It is this factor which determines the pressures and frictions, regardless of any effect due to increased wall thickness — all assuming the permissible increases are not exceeded.

The nature of the formwork face, particularly its permeability and surface condition, has a great bearing upon the friction and pressure factors. With timber forms (not the tongue-and-groove kind) gaps are left to allow for expansion. During the slide these are rarely closed, so they allow a certain amount of grout to pass through, thus reducing the lateral pressure and increasing friction. Treating the timber forms only proves effective for the first few metres of the slide, since the smaller particles become engrained in the timber surface, increasing, the friction and providing severe mechanical action.

Group two factors have extraordinary effects which are extremely difficult to portray numerically. This is because they are only calculable on the structure when all the relevant facts are known. However, four methods of calculating the effects have evolved.

Bohm's method

This revolves around the fact that pressures on the formwork are greater at low speeds. Calculations are therefore based upon a sliding speed of 10cm/hr and a separation point 600mm below the uppermost level of new concrete. Bohm also suggests a setting time of one hour, so producing a resultant force of 280kg/m for D acting 355mm down from the top (see Figure 3.2). A figure of 45kg/m for friction on the formwork sides is given, corresponding to the weight of a 250mm layer of new concrete 150mm wide.

Dreschel's method

In this case the earth-pressure theory is used, along with reduced hydrostatic pressure-distribution information and specific coefficients for the internal friction of fresh concrete. Hence for thick walls reduced hydrostatic distribution patterns are incorporated and for thin walls a silo pressure distribution is assumed. Drechsel assumes a maximum effective concrete head of 700mm and produces a D value of 485kg/m,

Figure 3.1 Section indicating formwork batter.

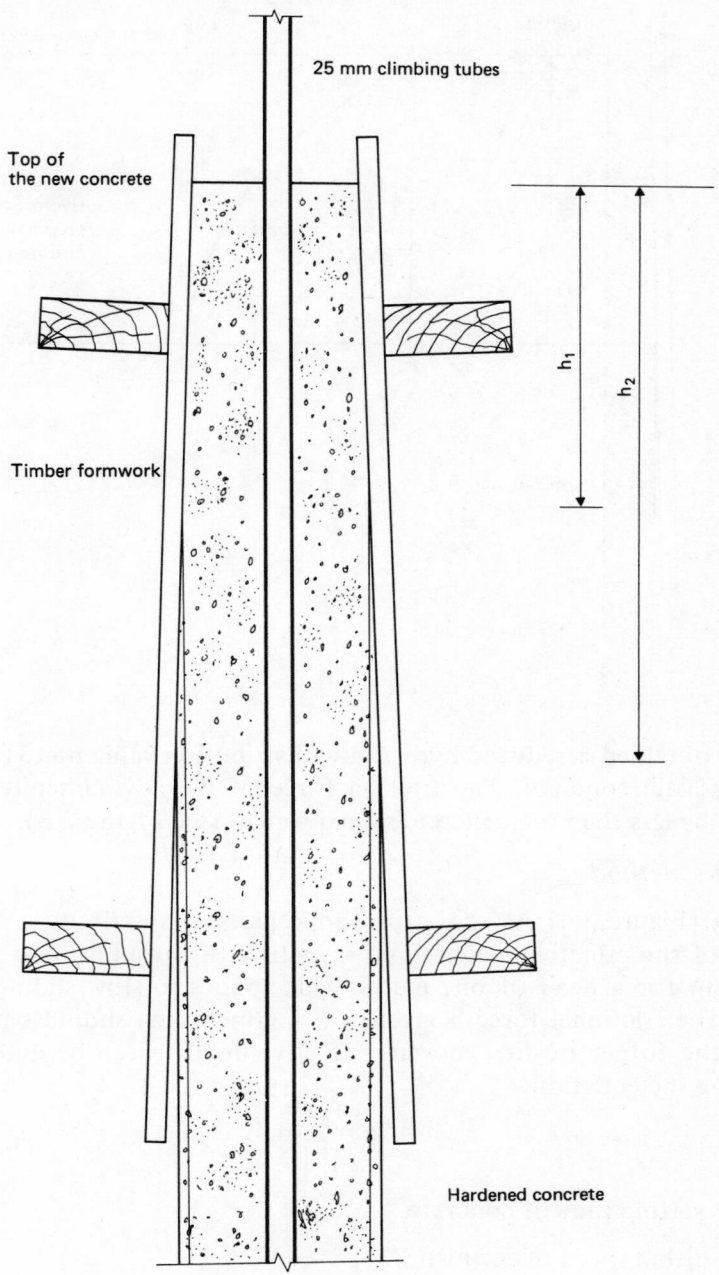

25 mm climbing tubes

Top of
the new concrete

Timber formwork

Hardened concrete

h_1

h_2

Figure 3.2 Formwork pressure distribution according to Bohm.

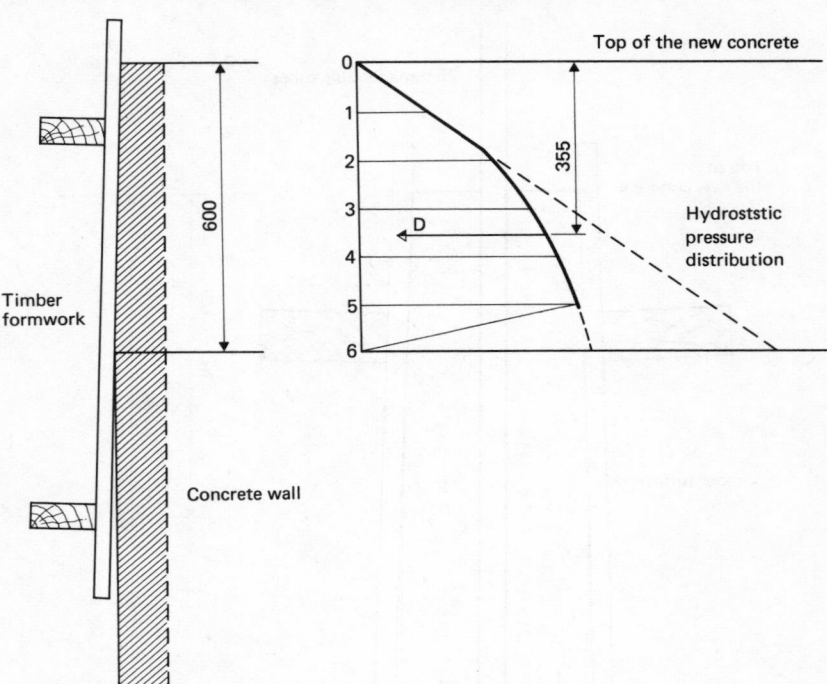

having obtained a reduced hydrostatic distribution value of 1310kg/m² for a plastic concrete. The friction force of the new concrete should always be less than the concrete's dead weight (see Figure 3.3).

Nennig's method

Nennig (Figure 3.4) assumes a parabolic pressure distribution over the depth of the effective concrete cross-section, giving a resultant force of 375kg/m for a head of one metre. This applies to slow sliding speeds only. The frictional force is given as 75kg/m, which should be trebled when the forms are first moved. Effective depth h can be determined by using the equation:

$$h = 2a = 2v_b t_v$$

where:

t_v = setting time of concrete,

v_b = rising speed of formwork,

a = ½h.

Figure 3.3 Formwork pressure distribution according to Drechsel.

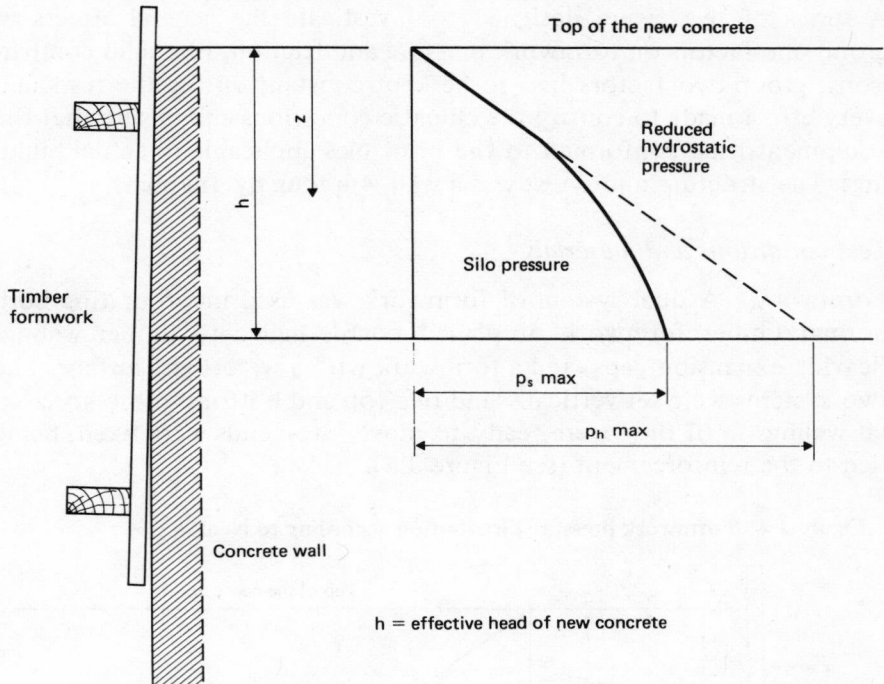

The resultant horizontal force per metre is therefore determined by the equation:

$$P_h = 2/3 \ (v_b t_v)2$$

American regulation

This gives a formula for the calculation of exerted pressure on sliding formwork and works on a speed of 40cm/hr, a concrete temperature of 40°C and a formwork height of 1200mm. The resultant force of 1,100kg/m was obtained by using the following formula:

$$P = C_1 + 6000 \ R/T$$

where:

C_1 = 100 (co-efficient of vibration),

R = speed of formwork in ft/hr,

T = temperature of concrete in °F,

P = lateral pressure in lb/ft²

Experimental work

A series of tests were designed to investigate the general effects of group-one factors on formwork pressure and friction. For valid comparisons, group-two factors had to be kept constant during the tests and every effort made to control the climatic conditions and ensure that the equipment used conformed to the principles applicable to actual buildings. The structure under test was a wall, 4m long by 4m high.

Test conditions and materials

Formwork. A dual system of formwork was used incorporating both normal timber formwork of planed boards nailed to timber walings (leaving expansion gaps) and a formwork with a watertight surface. The two systems were set vertically and tied top and bottom to the horizontal walings until they were ready to move. Stop ends were fixed, being tied to the reinforcement (see Figure 3.5).

Figure 3.4   Formwork pressure distribution according to Nennig.

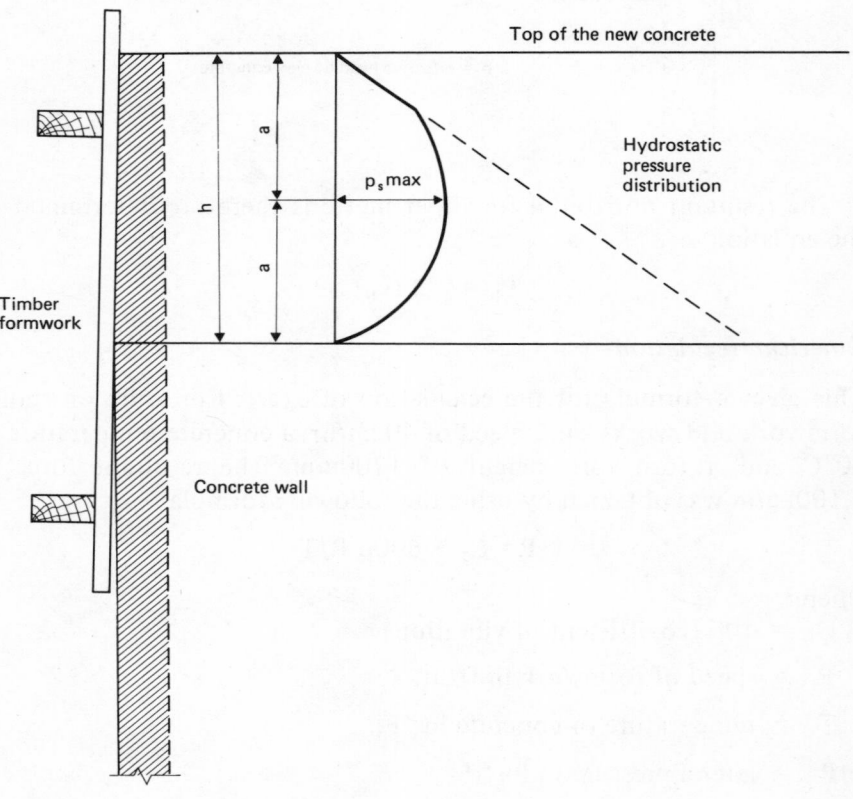

Figure 3.5 Test formwork arrangement.

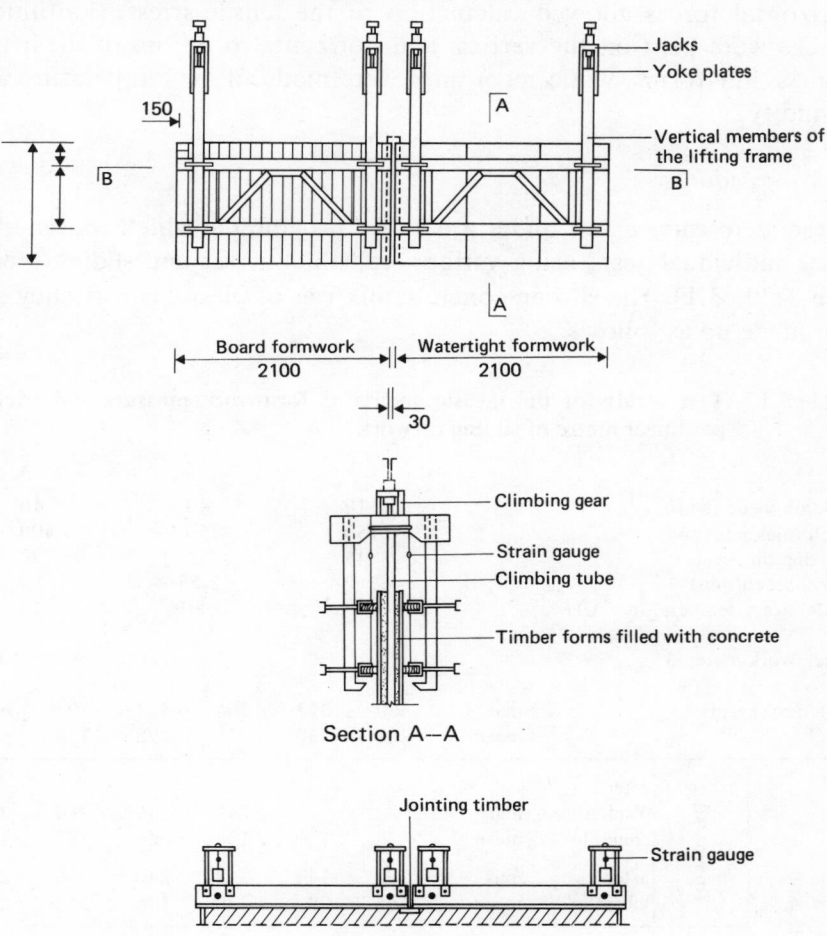

Section A--A

Section B--B

Reinforcement. For both types of formwork, a series of 10mm-diameter mild-steel bars were used, spaced at 200mm centres horizontally and vertically, in both faces.

Hydraulic equipment. Four sets of hydraulic jacks were used to lift the formwork, each connected by a ring main to the main pump.

Measurement of forces

The forces built up within this formwork system were determined by recording strain on duralumin strips coupled to strain resistance gauges and 3 or 8 loop oscillographs. All the forces transmitted to the supporting walings were pre-determined, so correct separation of vertical and horizontal forces allowed calculation of the tensile stress. Continuous checks were kept on the vertical and horizontal positions of the lifting frames and forms, while recordings were made of air temperature and humidity.

Test procedures

These were carried out under a detailed programme which consisted of three individual tests using various wall thicknesses and sliding speeds (see Table 3.1). The chosen concrete mix was of plastic consistency and was made up as follows:

Table 3.1 Test results of the measurements of formwork pressure and friction per linear metre of sliding forwork.

Test			1		2		3	
Sliding speed (cm/h)			10		40		40	
Wall thickness (mm)			150		150		400	
Setting time (min)			45		15		20	
Final set (h/min)			3:50		2:55		5:5	
Laboratory temperature ($^{\circ}$C)			19		20		14	
Formwork material			WT	BT	WT	BT	WT	BT
Friction (kg/m)		max	360	377	204	404	194	600
		mean	214	238	78	290	138	460
Pressure on the formwork (Kg/m)	Upper waling — after placing concrete	max	—	—	240	140	280	214
		mean	—	—	134	66	197	146
	Upper waling — after vibration	max	200	144	312	240	326	278
		mean	165	99	230	176	240	233
	Lower waling — after placing concrete	max	—	—	270	143	374	364
		mean	—	—	210	120	330	298
	Lower waling — after vibration	max	—	—	320	133	440	480
		mean	—	—	268	119	400	352
	Lower waling — during lifting	max	128	92	320	140	—	—
		mean	64	39	273	125	—	—
	resulting formwork pressure on the lifting frame	max	200	183	602	364	748	736
		mean	195	118	497	295	605	575

WT = watertight timber
BT = board formwork

46

8%	Finest particles	0 — 0.2
35%	Sand	0.2 — 3.0
17%	Sand	3.0 — 7.0
22%	Gravel	7.0 — 15.0
18%	Double broken chippings	15.0 — 30.0
350 kg/m	Portland cement giving a 28-day strength of 375 kg/cm^2.	

The test slide began three hours after the forms had been filled with concrete, which was placed in 200mm layers throughout the test and vibrated by internal pokers. On commencement of the slide the first set of measurements were taken, followed by others at 200mm intervals. Horizontal and vertical forces were only measured at three main stages in the sliding process, after placing the concrete; after vibration, and during formwork lifting. Maximum and minimum values were taken in each case.

Results

Figure 3.6 indicates the behaviour of the vertical forces once the pumps have been started. Vertical forces rise from D to E until form-work friction has been overcome. Once the formwork is sliding the forces decrease from E to F until the oil pressure drops to zero; where they further decrease to G is a result of slip in the lifting head. Any remaining vertical force is due to the weight of formwork, while the decrease in horizontal force during sliding is attributed to the inward batter of the forms. This force increases again while vibration is taking place at both sets of walings, in conjunction with the breakdown of internal friction; the extent of the breakdown is dependent on the vibrators' depth of action.

Evaluation of results

In Table 3.2 the maximum values obtained for formwork pressure and friction according to the four methods of calculation are compared with the experimentally determined maximum values. From this table it is possible to compare all the maximum values and it is apparent that the design load assumptions for the formwork pressure and friction values are derived from Bohm, Drechsel, Nennig and the American Regulation are inadequate. In general their stated values for pressure and friction are too low. With the progress of technical development and high sliding speeds becoming acceptable, increased formwork pressure and friction values are imminent as indicated by the sliding speeds used for the experiment.

Figure 3.6 Behaviour of the horizontal and vertical forces during measuring sequence.

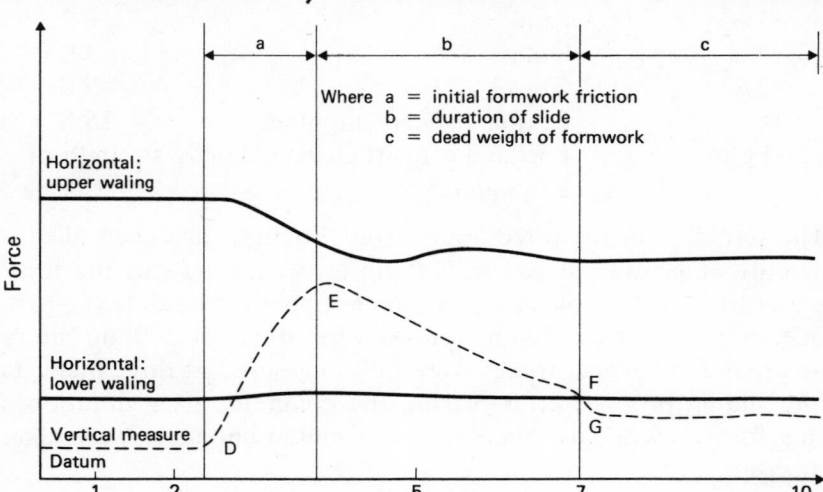

Formwork pressures

The effective head of concrete, which determines how much pressure is exerted on formwork, is affected by the depth of vibrator action, concrete setting time, sliding speeds and watertighness of the working faces.

Wall thickness, friction and choice of formwork are usually pre-determined by design. However, concrete compaction is not. It ranks high in the list of key considerations for good quality concrete and has to be carried out by an experienced operator. If the concrete is only lightly vibrated it will detach itself from the formwork when it is strong enough to carry its own weight and that of the above layers; thus the effective head is determined by initial set. On the other hand, with excessive vibration, the energy produced will seriously affect the already setting concrete deeper within the structure. This will produce unnecessary new pressures and may result in a weaker, less durable concrete. This problem may be avoided by incorporating a rapid-hardening cement in the mix, or by reducing the speed of the slide. Note too that if a watertight formwork is used, the trapped water transmits pressure to the formwork and affects the effective depth, while simultaneously reducing the initial strength of the concrete.

Looking at Table 3.3 and Figure 3.8, we see that the probable distribution of lateral pressure in concrete over the height of the formwork is determined by the horizontal forces, test 3 being used in this case.

Turning now to Figure 3.7, consider a metre-wide longitudinal strip and assume the formwork pressure does not change significantly. The distribution of concrete pressure $j(z)$ over the height Z of the formwork may then be taken as a linear load. The curve is hence defined by three limiting values:

(a) point A, which corresponds to the top of the freshly placed concrete and where $j(z) = 0$ because $z = 0$,

(b) hydrostatic pressure for the density of the freshly placed concrete $j(z) = pz$,

(c) point B, the point of concrete and formwork separation.

Vibration of the new concrete creates an acting hydrostatic distribution line AC. Lateral pressure exerted by the concrete between points C and B, however, decreases to zero at a rate related to the hardening process or the sliding formwork speed. The curve on the diagram may be determined from a step by step approximation evaluated from these results.

Known factors:

H_o and H_u, which are the values of the forces acting on the top and bottom walings.

To determine:

$P_o = H_o + H_u$

Z_o, which is the distance of P_o from the top of the new concrete, is evaluated by using:

$$Z_o = \frac{0.11H_o + 0.735H_u}{H_o + H_u}$$

For the first approximation of lateral pressure distribution related to formwork height, we will assume that:

hydrostatic pressure distribution $= (AC_1 B_1)$
resultant horizontal pressure $= P_1$ (assuming $P_1 = P_o$)

From these figures, $Z_1 = \sqrt{\dfrac{2P_o}{2.25p}}$

Once the values of Z_o and Z_1 have been determined for the measured values of H_o and H_u, it appears that Z_o is substantially less than Z_1, which leads to the assumption that any horizontal pressure is not equally distributed over the total height AB_1.

Table 3.2 Evaluation of the test results.

	Bohm (kg/m)	Drechsel (kg/m)	Nennig (kg/m)	American Regulation (kg/m)	Experimental results (kg/m)
Formwork pressure	280	458	375	1100	748 WT
Formwork friction	45	–	75	–	600 BT

WT = watertight formwork
BT = board formwork

Table 3.3 Mean value of twenty-eight readings related to Figure 3.8.

	Z_1 (mm)	P_1 (kg)	$P_1(z)$ (kg/m)
Plywood formwork	450	544	47
Board formwork	420	489	51

The second approximation is to assume that $P_2 = P_o = P_1$ and $z_2 = z_o$, as indicated by lines AC_2 D_2 B. The distance of C_2 D_2 and D_2 B_2 from the formwork and the top of the new concrete may be determined by:

$$F_1 = C_2 \ C_1 \ E_2 = F_2 = B_1 \ E_2 \ D_2 \ B_2 = F \text{ and } P_1 \ (z_1 - z_o) = F_a$$

where:

$$P_1 = P_o$$

$F = F_1 = F_2$ = forces acting at the area centroids

a = distance between centroids of areas F_1 and F_2.

Thus the lateral pressure will not change abruptly, but will present a curved distribution over the entire height of the formwork.

In the final assumption, indicated by the dotted curve, points B and C are connected by a curve shaped to allow the sum of forces F_4 and F_5 to equal zero ($F_4 + F_5 = F_3$).

In Figure 3.8 the average of 28 values obtained from the third test are plotted in accordance with the first approximation, then using the second approximation and lastly using the probable curve of the formwork pressure chart. The curve shows that the concrete detaches itself from the formwork at point B_2, 975mm from the formwork top. This indicates that the slide speed could have been increased, the limit being the position of the point B.

Thus, although the design principles of sliding formwork are universal, it is best to take as the adopted design load the formwork pressure related to point B. This is because trapezoidal pressure distribution gives an excellent approximation of the probable distribution of horizontal pressure and the resultant force position. These factors are crucial to the design of working faces and lifting frames. The horizontal boundary is determined by the formwork base and for the vertical, the boundary line distance from the formwork faces corresponds to half the pressure of the hydrostatic distribution 800mm below the top of the new concrete. The resultant pressure, 900 kg/m at 670mm from the top, gives a curve for horizontal pressure corresponding to the trapezoidal approximation. If a slow-setting concrete were required, the hori-

Figure 3.7 Pressure distribution and formwork depth.

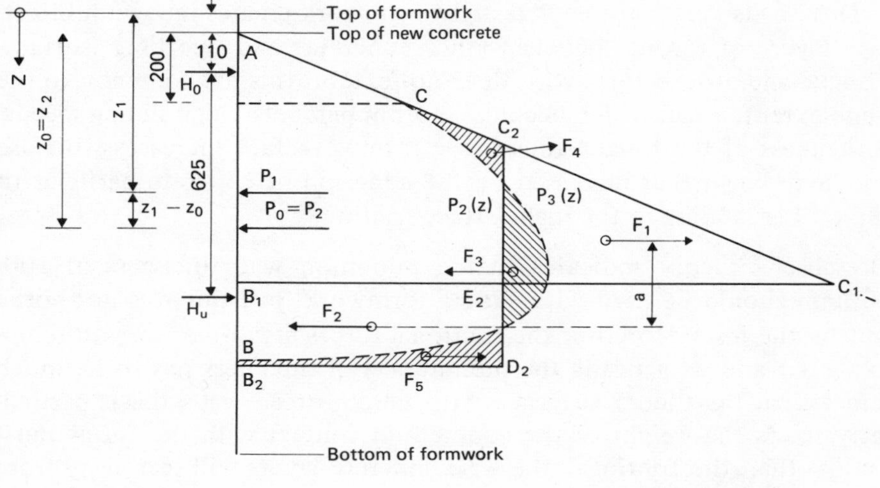

Figure 3.8 Maximum horizontal pressure on sliding formwork.

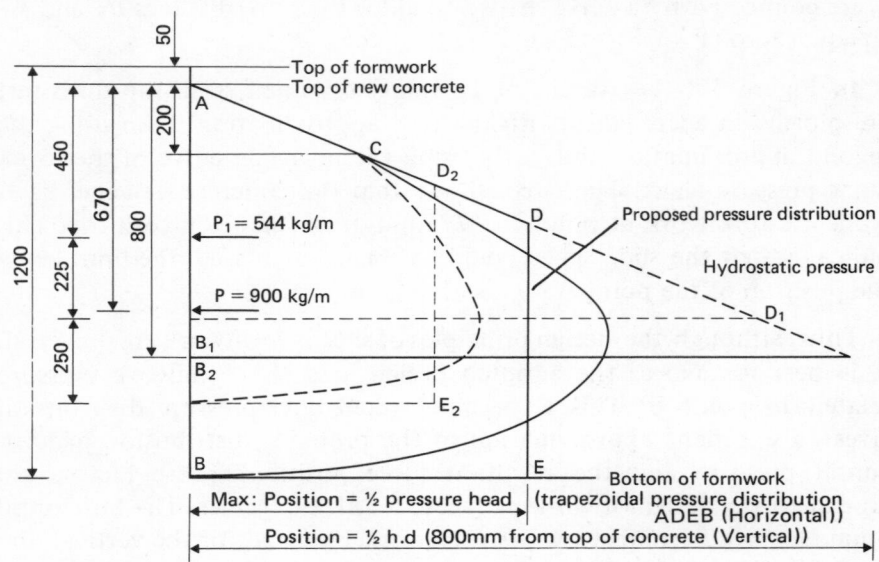

zontal pressure would increase and point B would descend. Slide speed would have to be reduced accordingly.

Formwork friction

Formwork friction depends on three main factors, each depending on the formwork facing material. These factors are the magnitude of the effective head on the formwork, the kind of working face and length of time between lifts.

Due to its very nature, watertight formwork creates its own lubricating layer of grout between the concrete and working surface. Tongue-and-groove formwork does not exhibit this phenomenon to the same extent, a fact which becomes more apparanet at high sliding speeds. Roughness of the tongue-and-groove formwork face increases with use and may be further increased by the adhesion of concrete particles to the timber, adding to the high surface friction.

Dreschsel's theory indicates that a minimum wall thickness of 100-200mm should be used with sliding formwork, but this was not borne out by the tests. Moreover the frictional forces measured were quite unexpected and as a result the minimum wall thickness has to be much larger than the theory suggests. With unreinforced walls this is particularly so, as the weight of the concrete in contact with the forms must be less than the friction; otherwise concrete layers will tear away from the wall and will attempt to stick to the formwork. Two-way reinforcement prevents this tendency and, as a bonus, transmits any forces to already hardened concrete.

Horizontal cracks may appear in the concrete when high sliding speeds or rapid-hardening cement are used. They arise due to a decrease in strength around the horizontal and vertical reinforcement connections, but they close up eventually under the weight of concrete above. With watertight formwork, of course, the flowing grout ensures that they are sealed more quickly.

Ideally, once the sliding formwork is in motion no long stoppages should occur as these will damage the texture of the concrete, especially near horizontal reinforcing bars. During the series of tests, a stoppage of 45 minutes produced spectacular results, with the boarded formwork creating a wide crack when restarted and the watertight formwork creating partial surface cracking.

Formwork batter

The magnitude of the batter depends upon four main factors: friction, formwork pressure, flexibility of the lifting frame and the number of connections within the lifting frame. A batter is always produced even if the formwork is set vertically at the start, due to pressure and friction' created when the system is moved.

Conclusions

Watertight formwork offers substantial advantages over the timber boarded system in that it provides a smooth, good quality surface finish. However, for a maximum resultant pressure of 900kg/m on the formwork, the maximum friction created by the boarded system was 750kg/m and only 400kg/m for the watertight system.

4 Slipform systems

During the last 40 years European and American companies have developed slipform into a method which is both practical and highly economic, assuming of course that the design lends itself to this form of construction.

Generally, the two main systems which are available to contractors in the UK both use a central power unit to control the hydraulic jacks, which in turn lift substantial steel yoke frames. The formwork, usually of steel, is attached to the main frames, together with the decks which provide working and storage space.

Siemcrete One system

With this method hydraulic jacks powered from a central position climb on 48mm-diameter steel tubes, the large diameter providing extra strength to the assembly, an important feature (see Figure 1.3). A three-deck structure is used.

The upper deck acts mainly as a storage area but is often used as a distribution deck and incorporates the vertical reinforcing-bar templates. Many items of small plant are stored at this level, to reduce congestion on the lower decks.

The central deck provides the main working area and is level with the top of the formwork.

The lower deck gives access to the concrete directly below the form-work, to enable finishes to be applied and blockouts exposed. This deck is supported by means of a hanging scaffold but is itself quite rigid.

Siemcrete Two system

Basically this is a simplified version of the Siemcrete One system, the only differences being mechanical (see Figure 4.1). The 48mm-diameter steel tube of system one is replaced with a 30mm-diameter solid rod, forfeiting some rigidity. However this is tolerable because with this assembly two decks are incorporated, the first called the main deck and the second being the lower deck as described before. The space available for working and storage is thus vastly reduced, but the sectional weight of the system is also reduced, which is an advantage in some circumstances.

Figure 4.1 Typical details of Siemcrete Two slipform system.

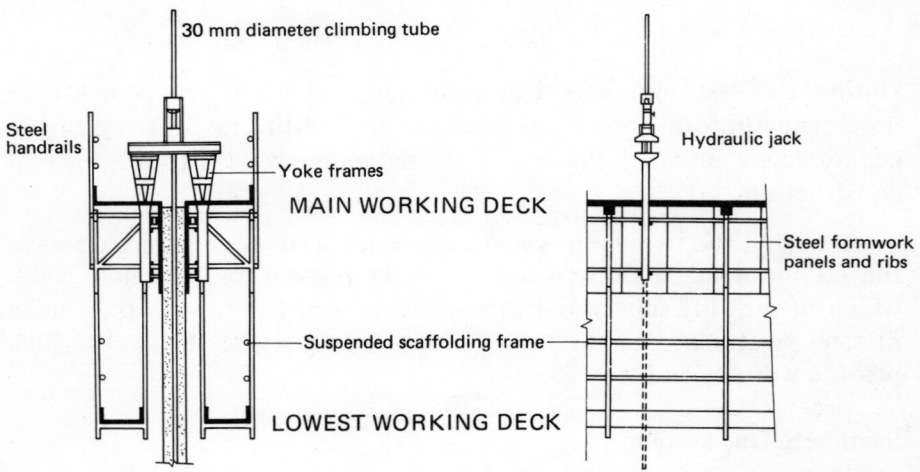

5 Case study

The contract covered by this narrative was undertaken in 1976 at Bolsover colliery, near Chesterfield, for the National Coal Board. The complete scheme included a coal storage area, magnet, washed coal crusher, screen house, transfer tower, 12 trestle bases and a 1000t cylindrical coal-storage bunker which made up the slipform contract (see Figure 5.1).

Designed by H. W. Rhodes & Partners of Nottingham, the bunker was intended to store coal transported from the production face directly to the uppermost part of the bunker. From there the coal flows down an internal open-spiral chute which eases further coal distribution to all outlets (see Figure 5.2).

This complex structure had a flat roof and two intermediate floors, the construction problems being aggravated by the inclusion of an internal inverted cone, to control coal flow. All these features had to be formed by traditional methods once the slide was complete.

Slipforming was done by British Lift Slab Ltd., which contracted to design, supply, erect, operate, dismantle and remove from site all the equipment needed to construct the 9.15m-diameter, 36m-high concrete bunker using the Siemcrete One system.

Naturally, services came high on the civil contractor's list of priorities. A hard access road was needed and material transport had to be arranged, including the provision of a crane. Also, suitable storage and working areas had to be set up. A 10kVA 440V supply (3-phase at 50Hz) was required for the slipforming equipment, along with a stand-by generator to ensure that the slide would not be interrupted by a power failure.

Furthermore, when the climbing tubes were exposed due to wall openings, it was the civil contractor's duty to supply adequate lateral

Figure 5.1 Case study: site plan of main contract.

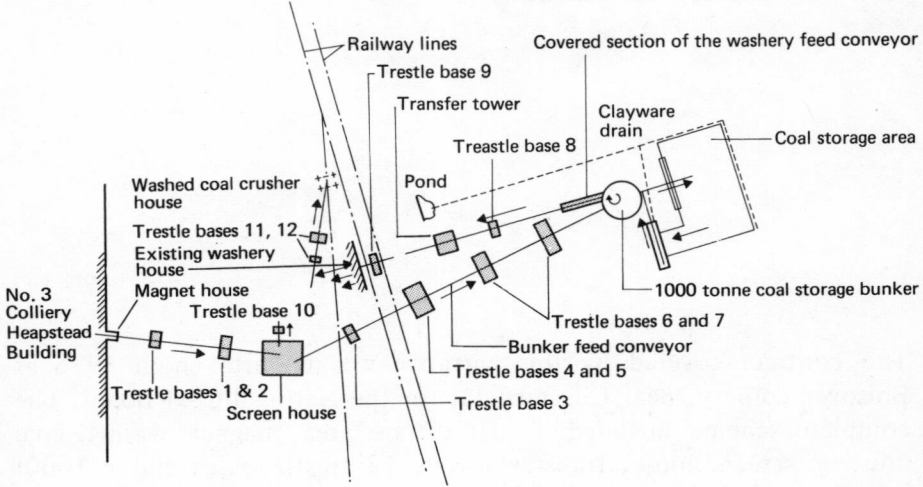

tubular bracing. It was also their duty to provide a means of access to the slipform structure, using ladders or power hoists. Weather protection trappings, concrete distribution chutes and platform lighting also had to be provided by the civil contractor. Platform lighting was to be quite separate from the main electrical supply system and needed its own standby generator. Moreover, statutory illumination requirements had to be satisfied.

Door, window and other wall openings were the responsibility of the slipforming team which had to ensure, for example, that the formwork could be secured to the reinforcement without moving horizontally or vertically. Unlike many construction methods, slipforming can successfully be commenced from a horizontal slab without any form of kicker. Steel reinforcement starter bars were incorporated in the slab to aid bonding and erection of the prefabricated forms.

Possibly the most controversial parts of the conditions of contract were the arrangements for providing joiners and labourers during the construction and dismantling of the slipforming equipment. These, it was stipulated, should be provided by the civil contractor to work directly under the slipform sub-contractor, reimbursing the main contractor for each man supplied.

The slide was supervised at all times by the sub-contractor's engineers, who undertook to complete the slide in a maximum of eight days and nights. This particular contract was unusual in that it incorporated two

Figure 5.2   Elevation of 1000t coal-storage bunker at Bolsover colliery.

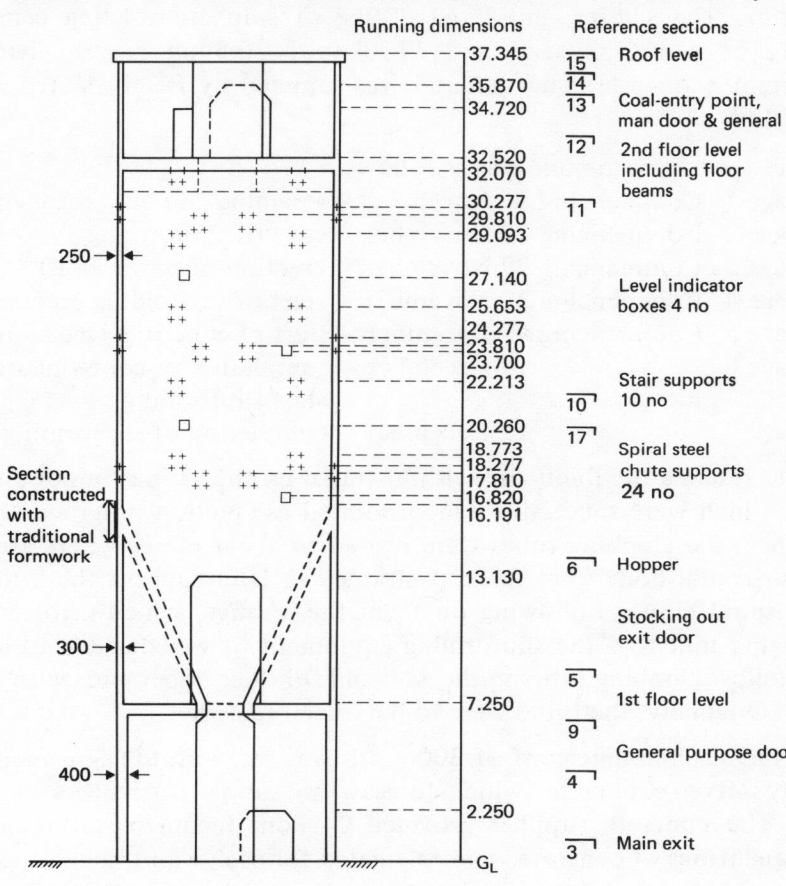

changes in wall thickness and a unique stage where the sliding formwork was lifted 2m without any concrete being poured. This tactic was dictated by the presence of huge numbers of reinforcing bars bonding the inverted cone to the main walls. Traditional formwork was used for this section, the concrete being poured through the steel sliding formwork into the timber formwork below (see Figure 5.3).

With the foundation work complete, British Lift Slab arrived on Monday November 15. The first stage was to collect suitable equipment and, particularly important, to explain the system, its equipment and basic principles to all the civil contractor's labour force and suppliers. Regular site meetings between the interested parties commenced on November 17 and included intricate discussions to decide, for instance, choice of concrete mix agreeable to designer, contractor and manufacturer. The chosen mix used 330kg of sulphate-resisting cement, 730kg of zone 2/3 sand and 1170kg of 20-5mm gravel, giving a 25N/mm² concrete. The material was supplied by Ready Mixed Concrete Ltd.

The agreed programme was as follows:

Stage 1 Commencing 15 November — assemble climbing formwork
Stage 2 Commencing 29 November — start of slipforming
Stage 3 Commencing 29 November — erection of power hoist
Stage 4 Commencing 30 November — start of scaffolding erection
Stage 5 Commencing 4 December — start of cone-ring (see above)
Stage 6   10 December — completion of cone-ring; restart slipforming
Stage 7   12 December — completion of slipforming

The sytem's flexibility is well illustrated by the various minor alterations which were successfully incorporated as erection progressed. For instance, the climbing tubes were resited to avoid places where groups of horizontal bolts were to be positioned to later support the internal steel spiral chute. Following on from this change, which involved rearranging much of the slipforming equipment, it was decided to leave the hollow climbing tubes in the wall and fill their upper ends with concrete. Originally, the tubes were to have been removed.

A target climbing speed of 300mm/h was set, with the agreement of Ready Mixed Concrete, which foresaw no supply difficulties at that rate. The concrete supplier provided 24 hour technician attendance. The quantities of concrete were calculated for each wall thickness as:

400mm-thick walls — 3.50m³/300mm height of wall
300mm-thick walls — 2.60m³/300mm height of wall
250mm-thick walls — 2.20m³/300mm height of wall.

Co-operation between the slipforming engineers and concrete technicians was of prime importance because of the prevailing cold weather. Cement content had to be corrected, additives certified, heaters used and slide speed slowed as necessary. Incidentally, in Scandinavia slipforming has been practised at temperatures as low as −18°C with extreme caution.

Throughout the initial week British Lift Slab's engineers were busy explaining to everyone on site how the system should operate. Meanwhile assembly of the sliding structure progressed, with the yoke legs, primary fixing of the sliding forms, timber decking arrangements and

Figure 5.3 Details of the traditionally constructed section of wall.

completion of steel reinforcement fixings all being done at low level (see Figure 5.4). To complete this stage a mobile crane was used and this remained in operation until the structure had grown beyond its reach, when a larger crane was substituted (see Figure 5.5).

The end of the second week's preparation saw both the upper and central decks complete, the hydraulic equipment all installed, all safety requirements complied with and a hired electrically operated hoist in operation, for which an engineer was supplied for the slide duration.

Even though the system was now operational, various details had to be sorted out before sliding could commence. Thus, although the lowest deck could not be fabricated until the slipforming apparatus had climbed to a suitable height, the suspended sections and decking units

Figure 5.4   View of the slipforming system under construction.

were made ready at ground level. Similarly, steel reinforcement, door frames and steel bolt-location plates were laid out in order of usage, each being suitably numbered and labelled.

Round-the-clock canteen arrangements were made to provide hot drinks throughout the two 12 hour shifts. To cope with the cold, tarpaulins, gloves, waterproof suits and portable gas fired heaters were supplied.

Good organisation of concrete distribution was clearly going to be important and the following system was evolved. The resident concrete technician would secure samples of each load of newly arrived concrete, making test cubes and performing slump tests before allowing it to be hoisted into position by skip and crane (see Figure 5.6).

Received by a nominated employee, who directed it to the central deck, concrete was tipped down a timber chute to waiting wheelbarrows. Alternate wagon loads were barrowed clockwise then anti-clockwise to prevent any rotation of the slipform apparatus (see Figures 5.7-5.9). Ordering, vibration and finishing of concrete were restricted to one person per operation per shift in order to eliminate error.

With so many openings and fixings in the walls, regular measurements were necessary, using a tape mounted at floor and slipform level (see

Figure 5.10). As the slipform apparatus climbed, the tape was automatically paid out. Measurements were painted on many yoke legs at
central deck level, giving heights of openings and fixings above floor
level. Each feature had its own colour code, an arrangement which
proved most valuable when used in conjunction with the tape.

Communications between the structure and ground level also
received considerable attention, with air horns mounted at each gate
being used to obtain the hoist driver's attention, while the crane driver
and slipforming structure personnel had two-way radios.

Considerable thought was given to the fixing system of the timber
formwork planned for the openings. This problem was greatly aggravated
by the structure being circular and the yokes being quite close together.

Figure 5.5 Crane change-over.

Figure 5.6 Typical concrete compressive strengths at 28 days.

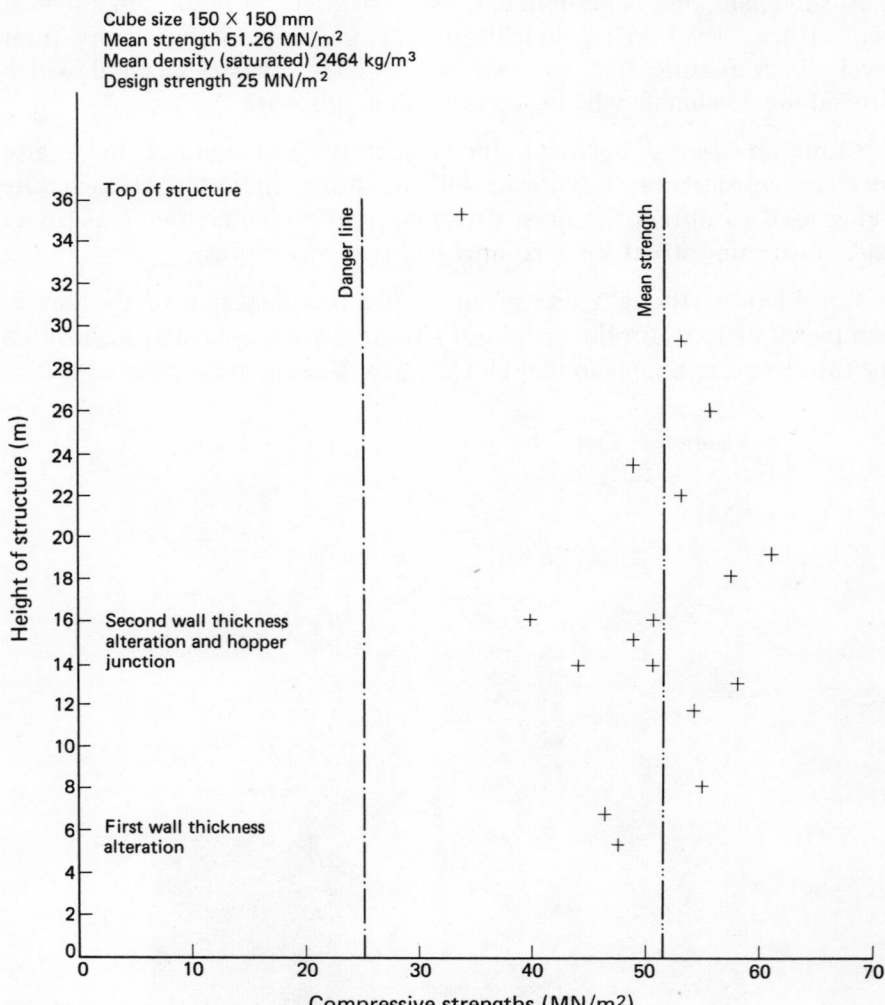

Cube size 150 × 150 mm
Mean strength 51.26 MN/m²
Mean density (saturated) 2464 kg/m³
Design strength 25 MN/m²

Formwork for the openings was manufactured in vertical sections not exceeding 400mm deep, allowing them to be fitted under the yoke cross-sections and shaped to suit the wall (see Figures 5.11–5.13). To prevent movement horizontally or vertically, the timber formwork sections had to be securely fixed without obstructing the sliding formwork panels. The steel bolt-location plates and sectional timber formwork frames were secured by tieing-in extra reinforcement with mild-steel binding wire (see Figure 5.14).

Figure 5.7 Unloading concrete on slipform's top deck. (Siemcrete One)

Figure 5.8   View of timber concrete chute from main working deck.

Figure 5.9 View of timber concrete chute from the top deck.

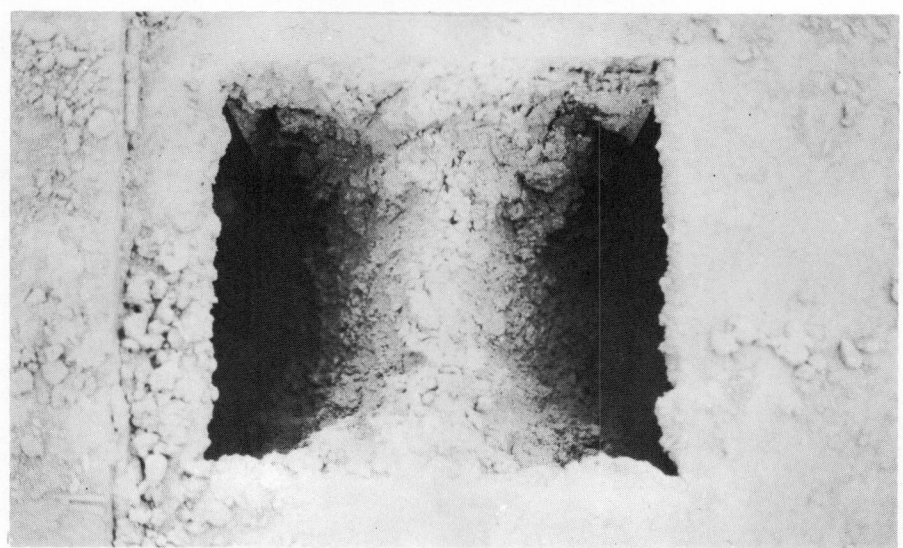

At this point it is worth explaining the function of these bolt-location plates. The design required a series of horizontal holding-down bolts to be cast within the wall, to house the internal steel spiral chute. With slipforming these groups of four bolts could easily be moved from their correct settings, so the designers specified that vertical steel plates, cut to the wall shape and with four horizontal tubes correctly spaced, should be fixed to suit each wall thickness. Known as the steel bolt-location plates, these were fitted with each tube sealed with tape and uncovered once the sliding structure had passed.

Sliding began as scheduled at 15.15 hours on November 29. By midnight the tape reading was 2.4m despite minor delays due to the need for a few further tests. The first of two wall-thickness reduction sequences was reached by 22.30 hours the following day but, unfortunately, changes in slide speed from 271mm/h to 141mm/h had been incurred due to steel-reinforcement fixing conditions. Thus the tape read only 6.8m. At this point the slipform equipment, void of concrete, was moved a short distance up the climbing tubes to facilitate fixing special L-shaped reinforcing bars, later used to tie the lowest concrete floor to the main walls. With these in place the slipform equipment was lowered onto them and onto the previously formed concrete wall, a practice normally unemployable as it tends to overwhelm the hydraulic jacks. (Figure 5.25 shows an accepted procedure for inserting starter

Figure 5.10   Mounted tape on the sliding equipment.

Figure 5.11　Timber formwork blockout sections, once the formwork has passed.

Figure 5.12  Timber formwork blockout section being fitted in place.

Figure 5.13 Height of yoke cross-section member above deck level.

Figure 5.14 Details of steel bolt location plates.

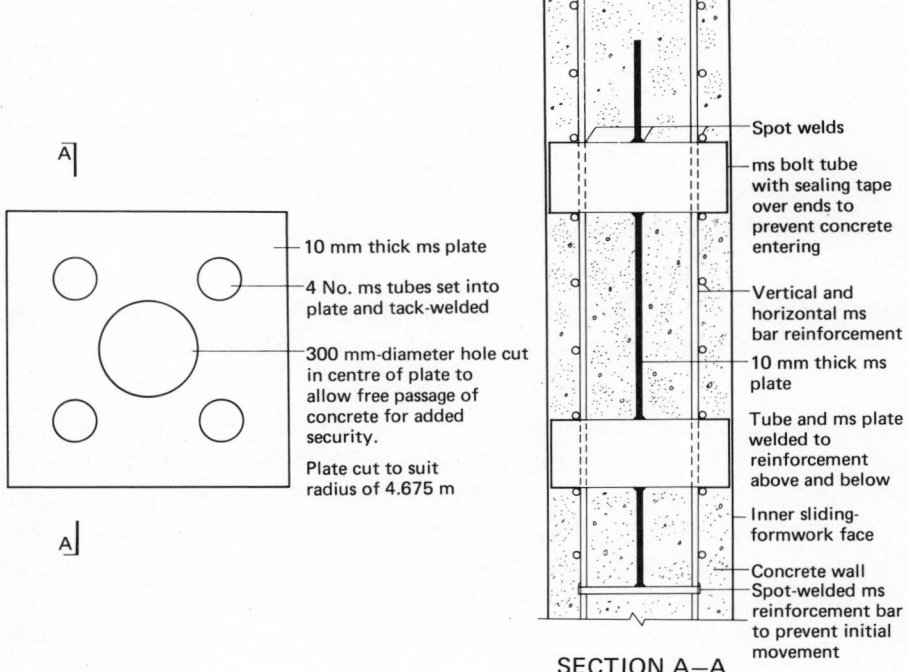

A|

— 10 mm thick ms plate

— 4 No. ms tubes set into plate and tack-welded

— 300 mm-diameter hole cut in centre of plate to allow free passage of concrete for added security.

Plate cut to suit radius of 4.675 m

A|

Spot welds

ms bolt tube with sealing tape over ends to prevent concrete entering

Vertical and horizontal ms bar reinforcement

10 mm thick ms plate

Tube and ms plate welded to reinforcement above and below

Inner sliding-formwork face

Concrete wall
Spot-welded ms reinforcement bar to prevent initial movement

SECTION A—A

Figure 5.15   Third deck positioned and polystyrene finish being applied.

bars although in this instance such a method was not used, due to the length of each horizontal member, their close proximity and general relation to each other.) Changing to a more slender wall section, by using a series of screw joints incorporated in the formwork design, proved to be fast and simple.

Climbing re-commenced at 6.30 hours on December 1, with all three working decks fully operational and a polystyrene finish being applied (see Figures 5.15 and 5.16). By midnight on December 2, sliding speed had again been reduced to provide extra time for the steel fixing personnel; nevertheless a tape reading of 13m was confirmed.

With two 12 hour shifts and ½ hour overlap at each shift change, competitive team spirit between the shifts became apparent. As the job progressed this became increasingly evident and was encouraged by the management.

On December 3 slipforming had to make way for traditional methods of shuttering, as the cone ring had been reached. No less than 696 reinforcing bars were incorporated within a vertical distance of 2m, most of them fixed at 45° to the main vertical walls (see Figure 5.17). By December 5 the sliding formwork had been moved up to facilitate traditional formwork methods. Such a procedure is most unusual and a great deal of bracing was applied to prevent movement of the slip-

Figure 5.16 View of a Siemcrete One system.

Figure 5.17 The 696 reinforcing bars for the cone ring.

form equipment which, in the absence of stabilising concrete, would otherwise have been a risk. Tubular scaffolding was erected both internally and externally to the bunker to provide a working platform for securing the heavy traditional formwork panels (see Figure 5.18). To make matters worse, the height of this platform was such that the slipforming system's lowest working deck had to be temporarily abandoned.

The traditional formwork, manufactured off site by Barcu Construction, consisted of softwood timber strips secured vertically to a framework of 125 by 50mm timber, which allowed the correct circular formation despite the large number of reinforcing bars. As this formwork system ended directly beneath the sliding apparatus, the ring could only be filled by passing concrete through the sliding formwork. British Lift Slab insisted the sliding formwork faces be protected from sticking concrete (which could spoil the surface finish) by lining the formwork with polythene sheets, which were removed before the slide restarted.

Meanwhile the slipforming apparatus was altered to accommodate the second reduction in wall thickness, a job which required a further 24 hours of industrious activity. The ring beam was completed on December 13 and the slide was restarted by 20.00 hours on December 14. Throughout the next 24 hours steady progress was maintained, confirmed by the night-shift tape reading of 23.55m.

Figure 5.18   Tubular scaffolding system.

Figure 5.19   Formation details of a door opening in a concrete wall.

Figure 5.20 Birds-eye view of a blockout stop-end.

During this stage of the operation a series of level-indicator boxes, door openings and steel bolt-location plates were incorporated within the wall. As previously discussed, forming the door openings presented some problems of formwork design, the results being shown in Figures 5.19 and 5.20. Fitting these sections at the correct intervals was critically important.

When internal and external walls are under construction simultaneously the whole of the central deck would be boarded out and used as a working, access and storage area (see Figures 5.21–5.23). Under such circumstances it is easy to overload the slipforming equipment by carrying too much material. This can affect the sliding action, so in such situations the number of jacks is often doubled, to provide a safer operating margin and also to allow rapid correction of any uneven

Figure 5.21 Main working deck level boarded over, Siemcrete One system.

Figure 5.22   Top-deck level boarded overs, Siemcrete One system.

Figure 5.23 Top-deck level boarded over, Siemcrete One system.

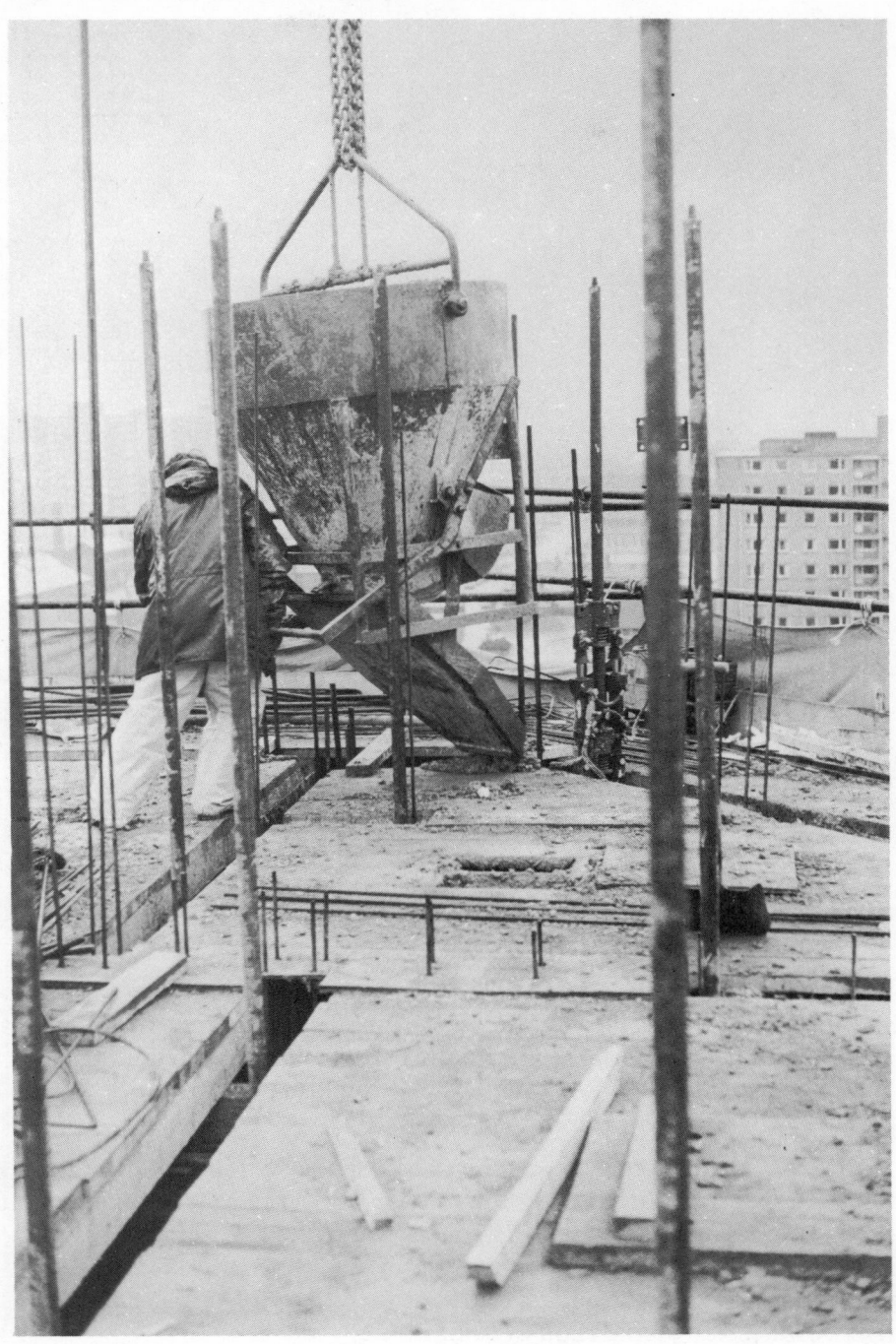

Figure 5.24 Case study nearing completion.

movement. In the case study, the working and storage deck loads were kept to a minimum.

By midnight on December 16 the slide speed had been reduced to 244mm/h which nevertheless produced a tape reading of 29.4m; a heavy frost had dictated the addition of more cement and a reduction in slide speed. Further heavy-duty tarpaulins were hung around the decks' perimeter (see Figure 5.24).

Throughout the contract each shift was responsible for making certain that materials were lifted onto the storage decks, and that wall inserts, timber frames and steel reinforcement were fixed as far as possible. Operatives actually working on the slipform decks were kept to a minimum, to avoid congestion, so everyone had specific duties allocated, right down to the preparation of hot drinks.

Figure 5.25 Typical arrangement for fixing starter bars in slipformed wall.

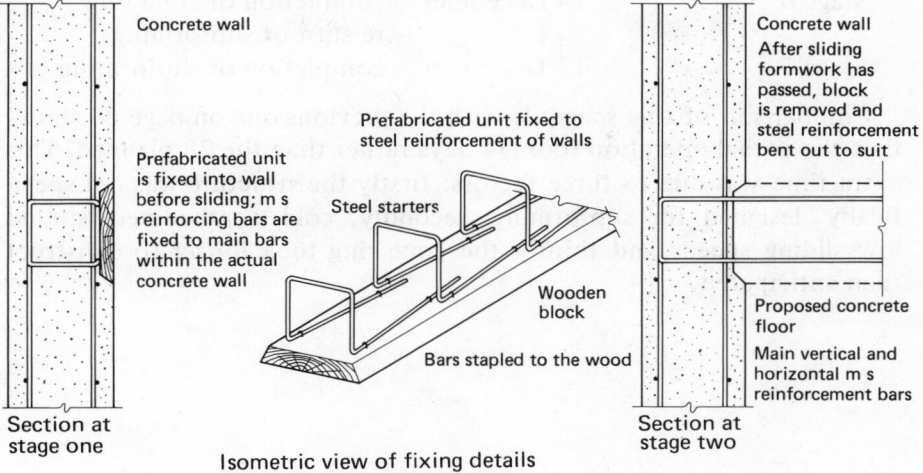

Concrete wall

Prefabricated unit is fixed into wall before sliding; m s reinforcing bars are fixed to main bars within the actual concrete wall

Prefabricated unit fixed to steel reinforcement of walls

Steel starters

Wooden block

Bars stapled to the wood

Section at stage one

Isometric view of fixing details

Concrete wall

After sliding formwork has passed, block is removed and steel reinforcement bent out to suit

Proposed concrete floor

Main vertical and horizontal m s reinforcement bars

Section at stage two

At the upper floor level floor-reinforcement tie bars had to be protected from sliding formwork (see Figure 5.25). Where the concrete floor support-beams joined the wall, polystyrene blockouts were secured to form a keyed recess which could be removed shortly after the formwork had passed. This enabled reinforcement to be located easily. Towards the end of the operation good progress was made and by 18.00 hours on December 18 sliding was complete.

With the required height reached, the equipment was taken further up until it cleared the concrete wall completely. This action ensured that the whole structure obtained a uniform surface texture and prevented concrete adhering to the formwork panels. Extra-long climbing tubes are incorporated to allow this action and are cut away after the system has been dismantled. Dismantling the slipforming equipment, using a hired crane, was accomplished by removing large portions of the system to ground level and only then reducing them to their component parts. Lasting two weeks, this undertaking required great attention to safety.

The slipforming contract was now complete. The actual timetable worked out as follows:

Stage 1 Commencing 15 November — assemble climbing formwork
Stage 2 Commencing 29 November — start of slipforming
Stage 3 Commencing 29 November — erection of power hoist
Stage 4 Commencing 30 November — start of scaffolding erection

Stage 5   Commencing   5 December — start of cone ring
Stage 6   14 December — completion of cone ring
                                    — re-start of slipforming
Stage 7   18 December — completion of slipforming

Comparison of this schedule with the previous one on page 60 shows that the actual operation took 34 days rather than the 28 planned. This extra time was due to three factors: firstly the structure was not specifically designed for slipforming, secondly, cold weather necessitated low sliding speeds and thirdly the cone ring took longer to construct than anticipated.

Appendix

Note 1

Figures A1–A5 relate to the case study and show how openings, steel bolt-location plates and level-indicator boxes were located prior to fixing. These diagrams highlight slipform's ability to accept large numbers of varied wall fittings and blockouts without undue concern. Note, the fixing rate and close proximity of many of the wall fixings, eg, Figure A.3 section 11–11.

Note 2

Plan details of the lowest slipform deck of the case study are shown by Figure A.6. This is the standard arrangement for a circular structure with no internal walls.

Note 3

Main working deck details of a typical slipform system are shown by Figure A.7, which clearly differentiates the concrete wall from the internal and external platforms. Note how the internal decking timbers are lapped at their joints in order to cater for increased deck areas when reductions in wall thickness are carried out.

Note 4

Figure A.8 provides plan details for the top slipform deck of the case study and shows deck support bearers, 6mm-diameter mild-steel spider and the relation of the steel sliding formwork's leading edge to the top deck.

Note 5

The relationship between the steel sliding formwork, steel rib sections, yoke frames, spider and hydraulic jack lifting points are all highlighted by Figure A.9 which provides the general arrangement details for a circular structure.

Note 6

In Figure A.10 the Siemcrete One system is featured, showing the top working deck plus details of hydraulic-jacks, yoke frames, steel reinforcement and climbing tubes.

Note 7

A continuation of Figure A.10, Figure A.11 reveals the sectional details of the Siemcrete One system below the hydraulic jacks' base line. Both the main working and lowest decks can be clearly identified.

Note 8

Figure A.12 clearly indicates the internal faces of the sliding formwork, steel reinforcement (both horizontal and vertical), the timber main working decks and the concrete wall.

Note 9

Considerable detail of the hydraulic jacks can be found in Figure A.13, which shows the jack extended ready to move the system upwards. The flexible tubes are hydraulic pipelines and a manual control valve is shown. Chalk marks scribed on the climbing tubes are used to level up the whole slipform system and can be especially useful on large-plan systems. An unusual point with this photograph is the jack's position: it is situated at the top working-deck level above a traditionally placed jack of the same type in a Siemcrete One system. The use of two jacks on a single climbing tube allows incorrect movements to be quickly remedied and heavy live loads imposed without any danger of overloading.

Note 10

The external working platform of the main working deck is illustrated in Figure A.14. Details which are clearly noticable include scaffolding-type handrails, toe boards, tarpaulin protective sheet, yoke frame legs, and timber blockout. The limited working space is also noticeable.

Note 11

From Figure A.15, it can be seen how close the yokes' cross-sectional bracing member is to the formwork's leading edge. This proximity permits only a few horizontal reinforcing bars to be fixed at a time.

Note 12

Figure A.16 indicates how a timber blockout section is fixed in place. Figure A.17 shows the timber formwork after the sliding equipment has passed. The external appearance seems quite rough, but can be treated from the lowest deck. In this case the concrete structure was to form the core of a new building, so finish was relatively unimportant.

Note 13

Figures A.18 and 19 are intended to illustrate the type of concrete finish which may be obtained using sliding formwork. The system moves in a series of jerks and Figure A.18 shows the result. It also illustrates the actual steel formwork panels in their working position. Figure A.19 shows the concrete surface once the movement lines of the formwork have been erased, which may be done by gently rubbing the surface with polystyrene biocks.

Note 14

Figure A.20 illustrates a convenient method of getting up and down the structure. In this instance the hoist gate is located on the top deck of a Siemcrete One system, but it may be sited on the main working deck if convenient.

Note 15

Figure A.21 is a view from inside a slipform operation, looking up to the main working deck's underside. Note how the spider is formed of steel channel sections interlaced between each formwork face. In this instance artificial lighting was provided because the main working deck was completely boarded over.

Figure A.1 Case study: sections relating to Figure 5.2.

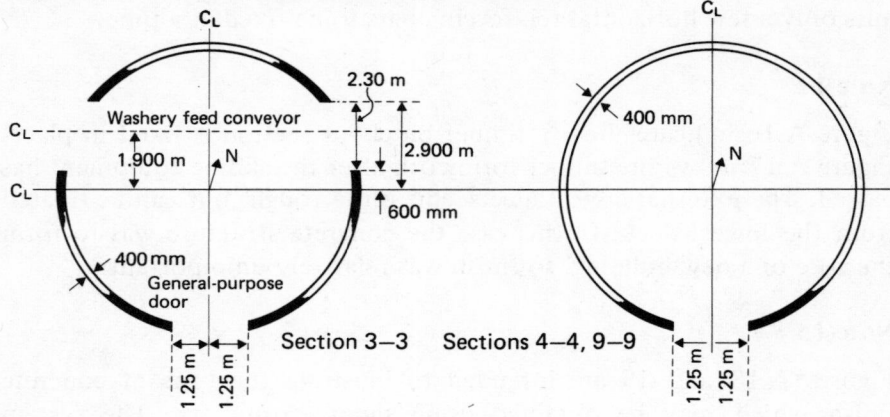

Section 3–3   Sections 4–4, 9–9

Figure A.2 Case study: sections relating to Figure 5.2.

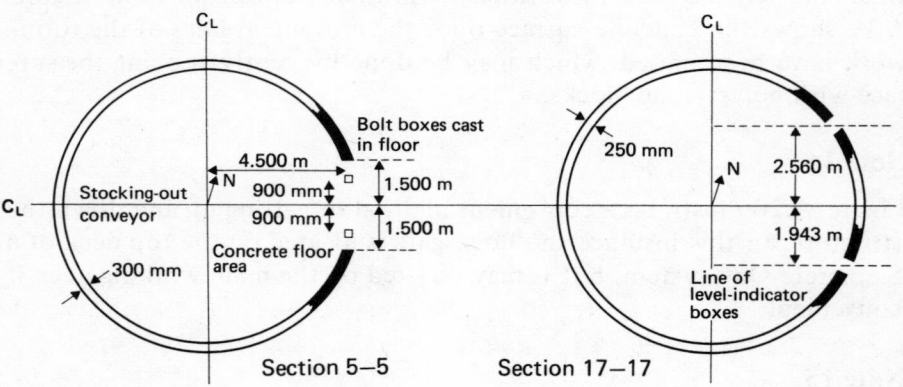

Section 5–5   Section 17–17

Figure A.3 Case study: sections relating to Figure 5.2.

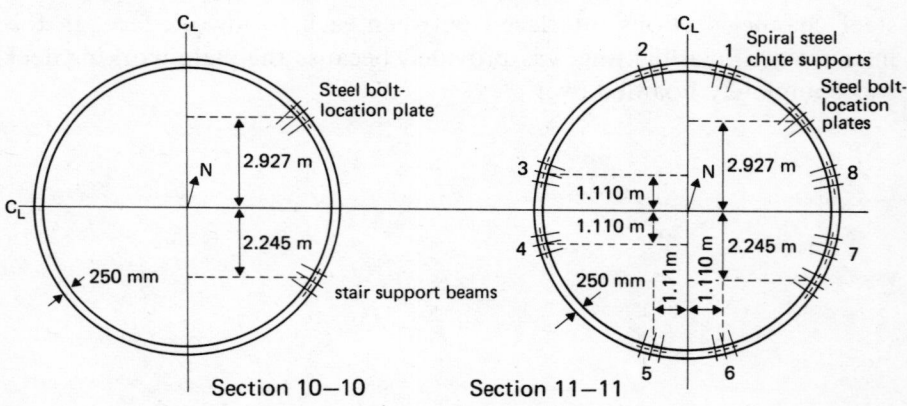

Section 10–10   Section 11–11

Figure A.4 Case study: sections relating to Figure 5.2.

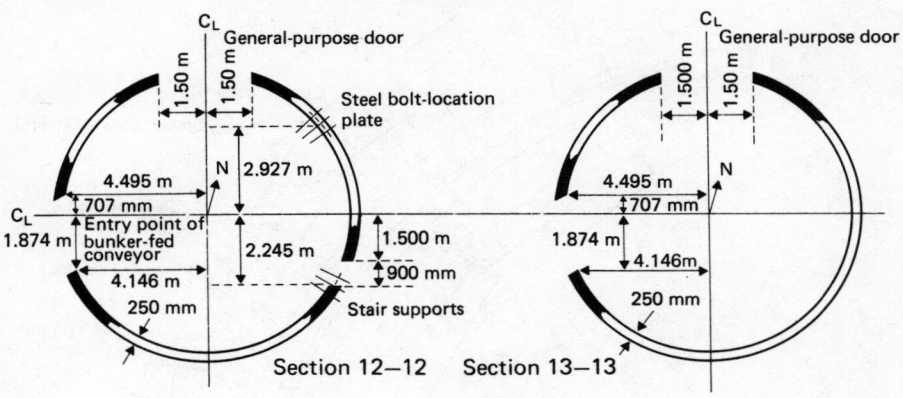

Section 12—12   Section 13—13

Figure A.5 Case study: sections relating to Figure 5.2.

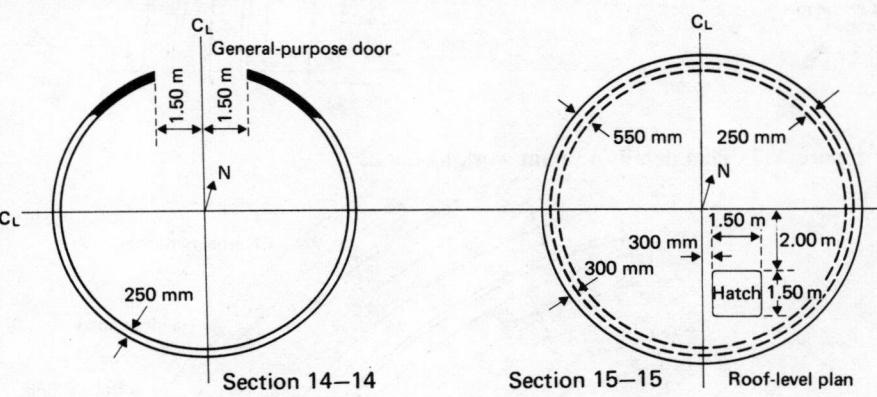

Section 14—14   Section 15—15   Roof-level plan

Figure A.6 Plan details of lowest slipform deck.

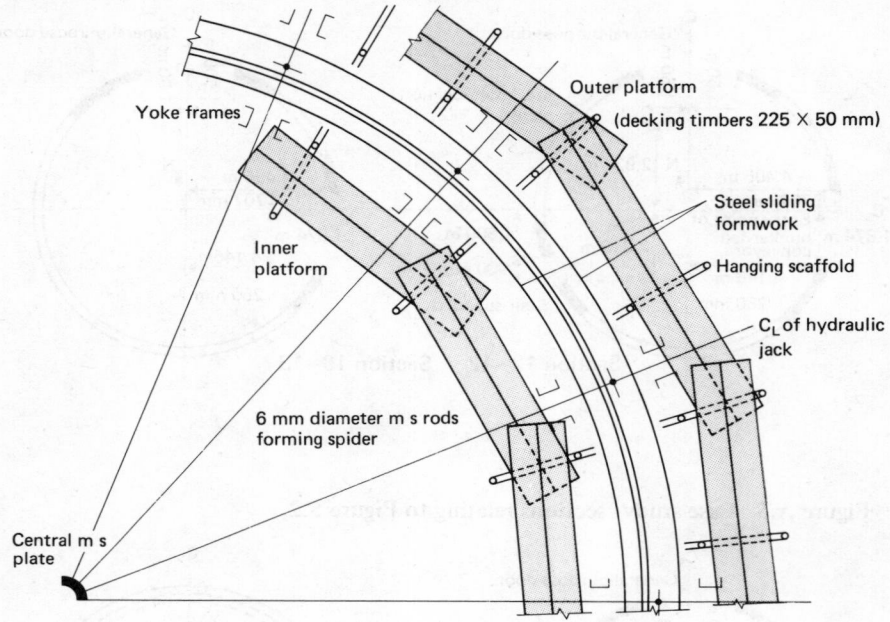

Yoke frames

Outer platform
(decking timbers 225 × 50 mm)

Inner platform

Steel sliding formwork

Hanging scaffold

C_L of hydraulic jack

6 mm diameter m s rods forming spider

Central m s plate

Figure A.7 Plan details of main working deck.

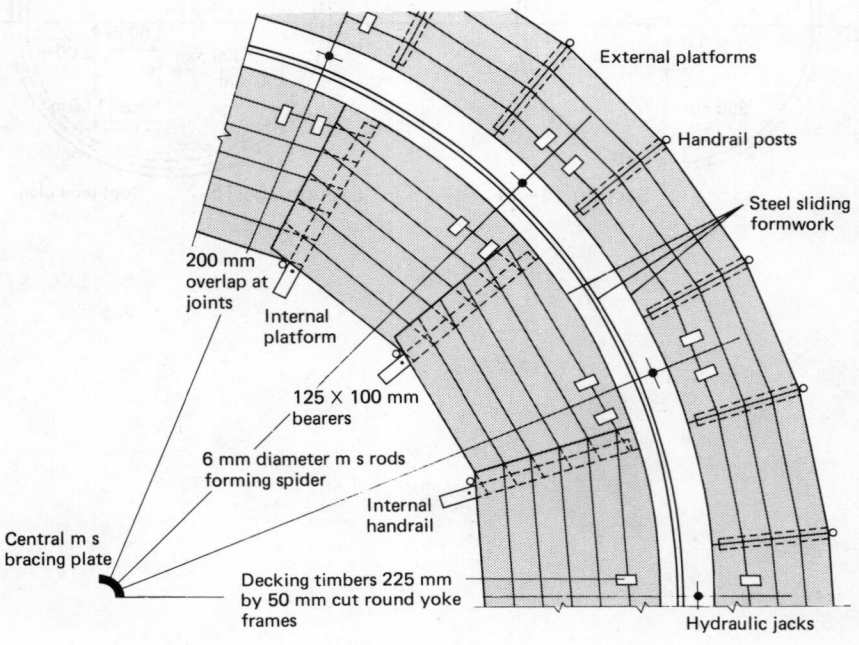

External platforms

Handrail posts

Steel sliding formwork

200 mm overlap at joints

Internal platform

125 × 100 mm bearers

6 mm diameter m s rods forming spider

Internal handrail

Central m s bracing plate

Decking timbers 225 mm by 50 mm cut round yoke frames

Hydraulic jacks

Figure A.8 Plan details of top slipform deck.

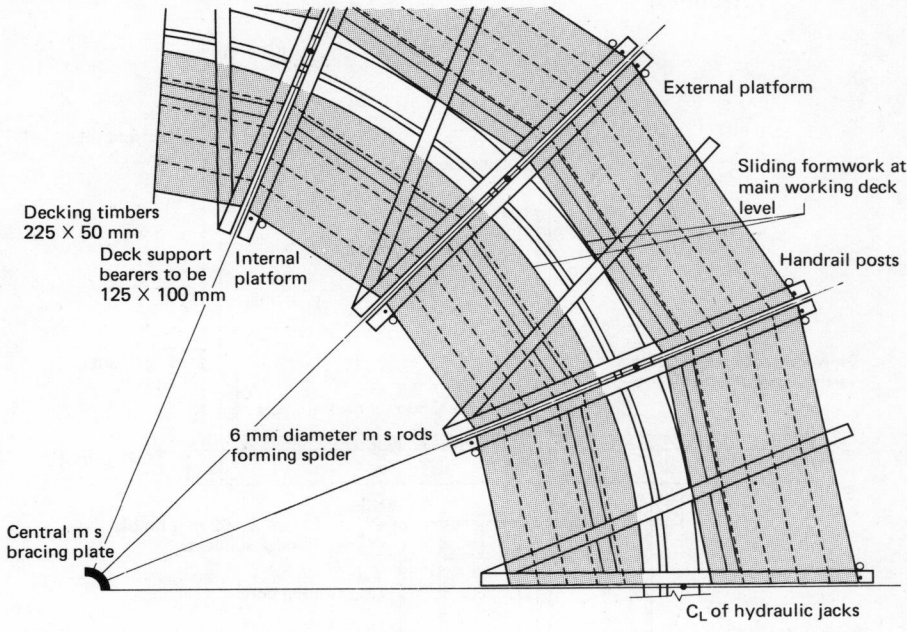

External platform

Sliding formwork at main working deck level

Decking timbers 225 × 50 mm

Deck support bearers to be 125 × 100 mm

Internal platform

Handrail posts

6 mm diameter m s rods forming spider

Central m s bracing plate

C_L of hydraulic jacks

Figure A.9 Plan details of sliding formwork.

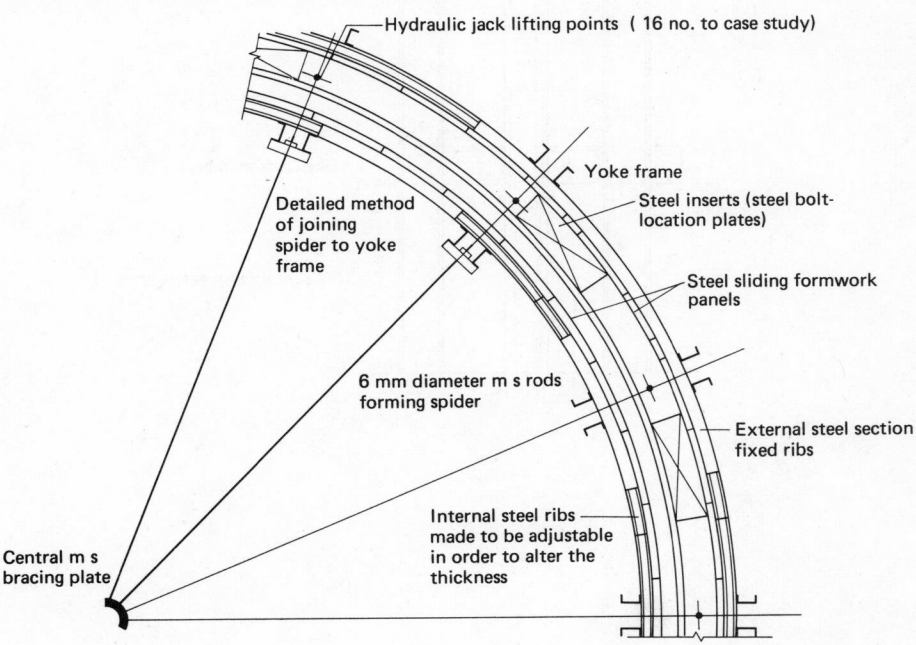

Hydraulic jack lifting points (16 no. to case study)

Detailed method of joining spider to yoke frame

Yoke frame

Steel inserts (steel bolt-location plates)

Steel sliding formwork panels

6 mm diameter m s rods forming spider

External steel section fixed ribs

Internal steel ribs made to be adjustable in order to alter the thickness

Central m s bracing plate

Figure A.10 Typical section through a slipform system indicating uppermost features.

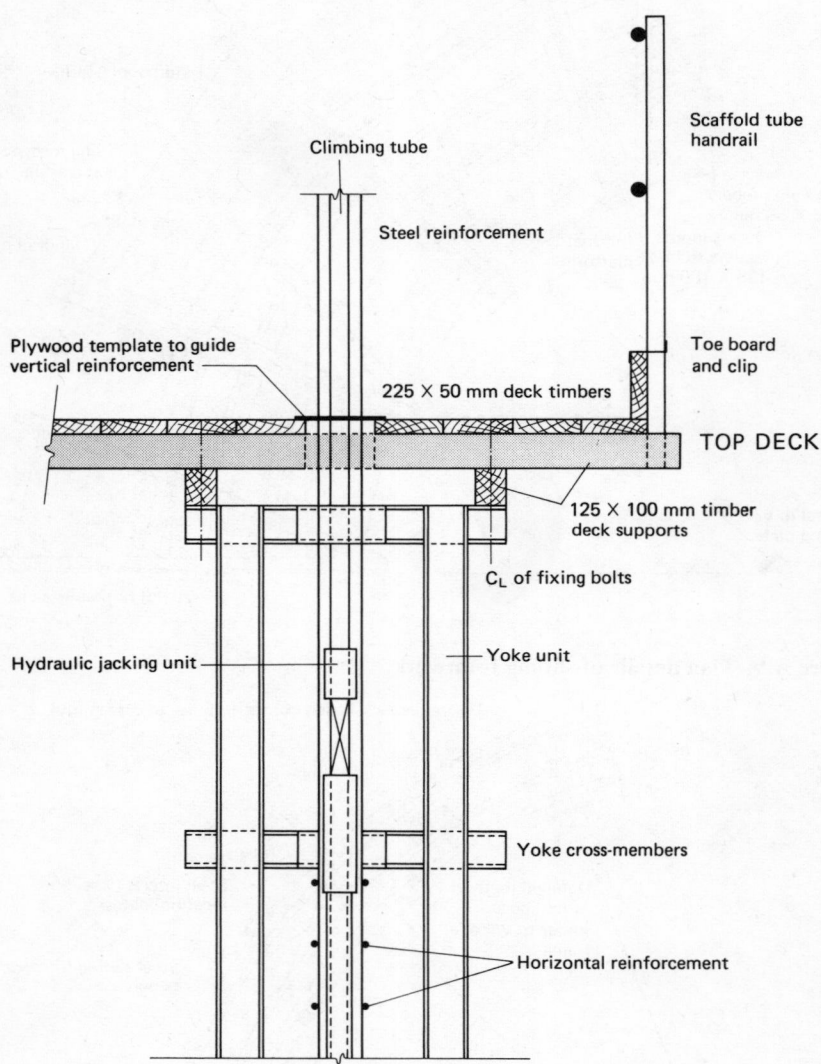

Climbing tube

Scaffold tube handrail

Steel reinforcement

Plywood template to guide vertical reinforcement

225 × 50 mm deck timbers

Toe board and clip

TOP DECK

125 × 100 mm timber deck supports

C_L of fixing bolts

Hydraulic jacking unit

Yoke unit

Yoke cross-members

Horizontal reinforcement

Figure A.11 Typical section through a slipform system indicating lower features.

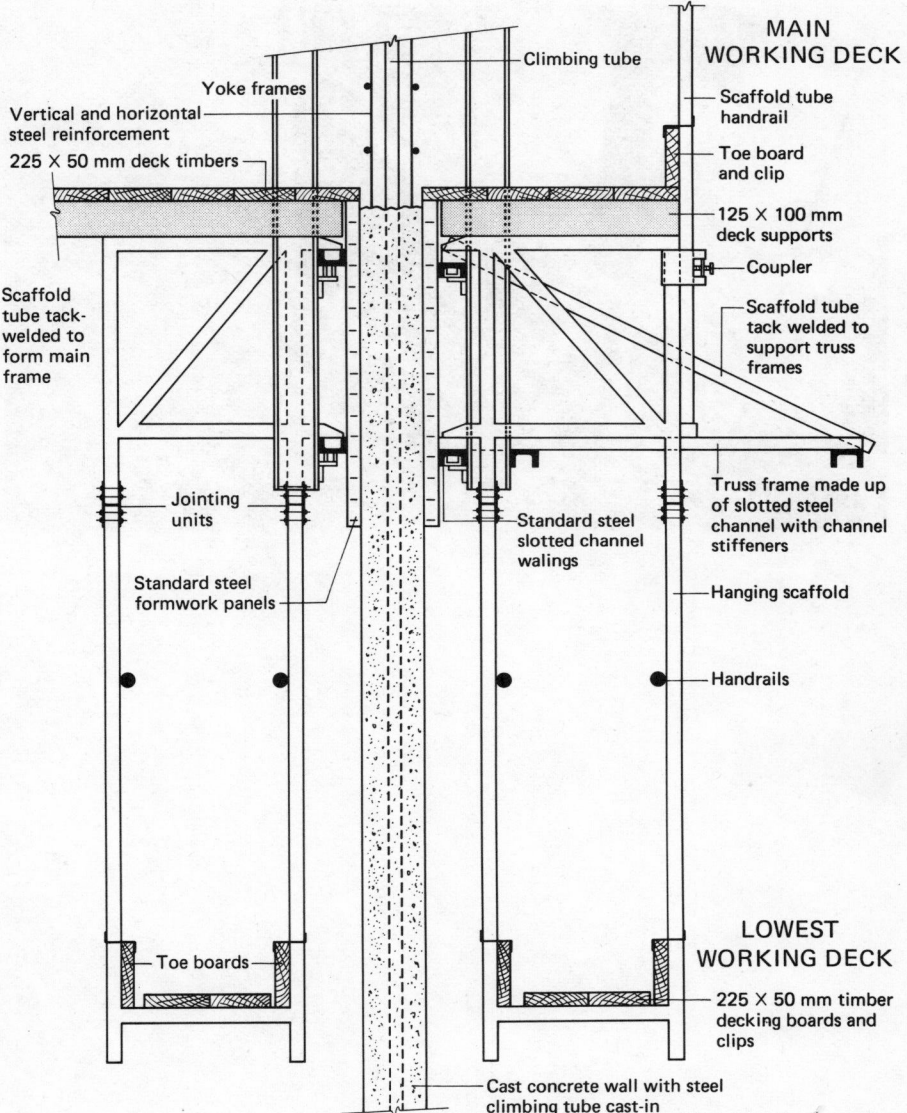

MAIN
WORKING DECK

Climbing tube

Yoke frames

Vertical and horizontal
steel reinforcement

225 × 50 mm deck timbers

Scaffold tube
handrail

Toe board
and clip

125 × 100 mm
deck supports

Coupler

Scaffold
tube tack-
welded to
form main
frame

Scaffold tube
tack welded to
support truss
frames

Jointing
units

Truss frame made up
of slotted steel
channel with channel
stiffeners

Standard steel
slotted channel
walings

Standard steel
formwork panels

Hanging scaffold

Handrails

LOWEST
WORKING DECK

Toe boards

225 × 50 mm timber
decking boards and
clips

Cast concrete wall with steel
climbing tube cast-in

Figure A.12 Sliding formwork's leading edge.

Figure A.13   Illustrated hydraulic jack.

Figure A.14   External working platform of the main working deck.

Figure A.15 Yoke frame and leading formwork edge/deck relationship.

Figure A.16   Timber blockout being fitted to steel reinforcement.

Figure A.17   Close-up of wall and blockout.

Figure A.18 Steel sliding formwork panels in-situ.

Figure A.19 Available concrete finishes.

Figure A.20 Power hoist and slipform relationship.

Figure A.21 View of the spider layout.

Bibliography

Barnbrook, G, "Concrete practice", Cement & Concrete Association, 1975.

Berry, J R, "Concrete round the clock", Concrete, October, 1974.

British Lift Slab Ltd, "Lift Slab, Slipform 18 Eq D6", British Lift Slab.

Chudley, R, "Construction technology volume 2", Longman.

Foster, J S, and Harington, R, "Structure and fabric, part 2", Anchor Press, 1976.

McSweeney, M F, and Grough, M E, "Central Bank offices, Dublin", The Institution of Engineers of Ireland, 1977.

Peurifoy, R L, "Formwork for concrete structures", McGraw-Hill, 1964.

Shirley, D E, "Introduction to concrete", Cement & Concrete Association, 1975.

Steinecke, M, "Experimental investigation of the pressure exerted on and the friction encountered by sliding formwork", Cement & Concrete Association, 1968.

Wynn, A E, "Design and construction of formwork for concrete structures; fifth edition", Concrete Publications Ltd, 1965.

"Slipforming complete at new Stock Exchange", Civil Engineering and Public Works Review, March 1968.

"Slipforming a tapered chimney in Denmark", Contract Journal, May 1968.

Index

glass fibre panels, 14

history, 9, 11
hoist, 61, 85
hydraulic pump, 20
hydrostatic distribution patterns, 40

jacks, 17, 31, 45, 76
   central hydraulic pump, 20
   early screw, 19
   electric motors, 20
   hollow screw, 20
   hydraulic, 20, 84
      overloading of, 19, 84
      pipelines, 84
   lifting capacity, 19
   movement, 80, 84
   operation of, 17
   pneumatic, 17
   water level, 29
joiners, 58

labourers, 58
levelling, 12
lifting operation, 12
lighting, 58, 85
London Stock Exchange, 29

manually controlled systems, 9
material supply, co-ordination of, 29
measurements, 62, 63

Nennig's method, 42, 47

openings, 63
   door, 76
optical
   instruments, 30
   plumbs, 28, 31
organisation, 11
origin, 9

polystyrene
   blockouts, 81, 83
   finish, 72
pressure, 40
   calculation of, 39
   distribution
      parabolic, 42
      silo, 40

   exerted, 43
   factors, 39
   lateral, 39, 48

Reading Town Centre, 27
reinforcing-bars, 59, 85
   templates, 55
ribs, 14, 27, 83
roof formwork, 17
rotation of slipform system, 11, 62

scaffolding, 74, 84
screw-jack method, 11
services, 57
shift system, 72, 80
shuttering, traditional, 72
Siemcrete One system, 30, 55, 57, 84
Siemcrete Two system, 55
silo pressure distribution, 40
site meetings, 59
slide speed, 28, 30, 42, 60, 67, 80
slipforming
   acceptability of, 9, 13
   advantages of, 24
   disadvantages of, 25
   equipment, overloading of, 76
   flexibility of, 60
   speed of, 28
      stoppages, 53
   traditional methods of, 57
slipform structure, acess to, 58
spiders, 16, 23, 83, 84, 85
stability, 11, 12
starter bars, 67
steel
   fixers, 72
   panels, 14, 85
   tapes, 31
   tubes, 55
stresses, 19
supervision, 58
supporting joists, 16
surface finish, 74, 81, 85

target speed, 31
technical ignorance, 13
thrust timbers, 16
tie bars, 81
timber blockouts, 84, 85
timber chutes, 62

RELATED TITLES FROM CONSTRUCTION PRESS

Testing and Test methods of Fibre Cement Composites, *edited by R N Swamy*
A4 550pp Hardcover 0 904406 94 6

Polymers in Concrete, *The Concrete Society*
A4 460pp Hardcover 0 904406 20 2

New Concrete Technologies and Building Design, *edited by A M Neville and M Chatterton*
A4 134 pp Hardcover 0 86095 846 9

Concrete Admixtures: Use and Applications, *edited by M R Rixom*
A4 88pp Hardcover 0 904406 32 6

Concrete, *Building Research Establishment Research Papers*
A4 340pp Hardcover 0 904406 37 7

Fibre Reinforced Materials, *Building Research Establishment Research Papers*
A4 168pp Hardcover 0 904406 38 5

CEB Manuals of Design and Technology

Autoclaved Aerated Concrete, *edited by G Bave*
A4 90pp Hardcover 0 904406 76 8

Buckling and Instability, *edited by A A Jakobsen*
A4 136pp Hardcover 0 904406 86 5

Lightweight Aggregate Concrete, *edited by A Short*
A4 178pp Hardcover 0 904406 24 5